T0387000

SFPE Guide to Human Behavior in Fire

Society of Fire Protection Engineers

SFPE Guide to Human Behavior in Fire

Second Edition

Society of Fire Protection Engineers
Gaithersburg, Maryland, USA

ISBN 978-3-319-94696-2   ISBN 978-3-319-94697-9   (eBook)
https://doi.org/10.1007/978-3-319-94697-9

Library of Congress Control Number: 2018956312

© Society of Fire Protection Engineers 2019
1st Edition: © SFPE 2003
This work is subject to copyright. All rights are reserved by the Publisher, whether the whole or part of the material is
concerned, specifically the rights of translation, reprinting, reuse of illustrations, recitation, broadcasting, reproduction
on microfilms or in any other physical way, and transmission or information storage and retrieval, electronic adaptation,
computer software, or by similar or dissimilar methodology now known or hereafter developed.
The use of general descriptive names, registered names, trademarks, service marks, etc. in this publication does not
imply, even in the absence of a specific statement, that such names are exempt from the relevant protective laws and
regulations and therefore free for general use.
The publisher, the authors, and the editors are safe to assume that the advice and information in this book are believed to
be true and accurate at the date of publication. Neither the publisher nor the authors or the editors give a warranty,
express or implied, with respect to the material contained herein or for any errors or omissions that may have been made.
The publisher remains neutral with regard to jurisdictional claims in published maps and institutional affiliations.

This Springer imprint is published by the registered company Springer Nature Switzerland AG
The registered company address is: Gewerbestrasse 11, 6330 Cham, Switzerland

SFPE Task Group on Human Behavior in Fire

CHAIR
Daniel J. O'Connor, P.E., FSFPE
JENSEN HUGHES

MEMBERS

Justin Biller, P.E.
Emerson Graham + Associates

Steve Dryden, P.E.
GHD

Rita Fahy, Ph.D.
NFPA

Richard G. Gann, Ph.D.
(NIST, Retired)

Simon Goodhead, P.E., CEng
JENSEN HUGHES

Norman Groner, Ph.D.
John Jay College, City University
of New York

Steve Gwynne, Ph.D.
National Research Council Canada

Bryan Hoskins, P.E., Ph.D.
Oklahoma State University

Michael Kinsey, Ph.D.
Arup

Erica D. Kuligowski, Ph.D.
NIST

Jamie McAllister, P.E., Ph.D.
NIST

Hidemi Omori
JENSEN HUGHES

David A. Purser, Ph.D.
Hartford Environmental Research

Steven Strege, P.E.
JENSEN HUGHES

STAFF
Chris Jelenewicz, P.E., FSFPE
SFPE

In the development of new areas of academic study, there is a need for individuals with commitment and passion that lead the way on research and studies that may seem quite academic and sometimes of little or no practical use. This Guide to Human Behavior in Fire is an exception in that the practical use and application of the Guide can in large part be attributed to the works, leadership, and mentoring provided by three originating members of the SFPE Task Group on Human Behavior in Fire.

This Guide is dedicated to the memory of three individuals who shared their knowledge and unique expertise in the development of the first edition of the Guide. Guylène Proulx, Ph.D., Harold E. Nelson, P.E., and John L. Bryan, Ed. D., made significant and internationally recognized contributions to the study and practical knowledge base that exist today for fire protection engineers. Their contribution to the study of human behavior was not only their work on the guide, but most importantly the awareness they raised in the study of human behavior and their influence on students, colleagues, fire protection engineers, and others interested in understanding and addressing the nuances of human behavior in fire situations. Many of those individuals influenced by Proulx, Nelson, and Bryan are now continuing the effort of further research and development of the methods and practices needed to better address the issue of human behavior. This second edition of the Guide is a tangible acknowledgment of their legacy contributions to the profession of fire protection engineering and fire-related human behavior.

Acknowledgements

The SFPE Task Group on Human Behavior in Fire would like to thank the following individuals who assisted in the development of this guide: Alberto Alvarez Rodriguez, JENSEN HUGHES; Karl Fridolf, WSP; Julie Bryant, UL Firefighter Safety Research Institute; and Madison West, SFPE.

Contents

1	**Introduction**	1
2	**Integrating Human Behavior Factors into Design**	3
	2.1 Use of this Guide	3
	2.2 Human Behaviour Assumptions Within Fire Codes and Standards	5
	2.3 Performance-based Design and Human Behavior Considerations Worldwide	6
	2.4 Time as Function of Behavior	8
	2.5 Informing the Engineer for Improved Consideration of Human Behavior	11

Part I Understanding Human Behavior in Fires

3	**Population Characteristics**	15
	3.1 Introduction	15
	3.2 Population Numbers and Density	15
	3.3 Alone Or with Others	16
	3.4 Familiarity with the Building	16
	3.5 Distribution and Activities	16
	3.6 Alertness	17
	3.7 Physical and Cognitive Ability	17
	3.8 Social Affiliation	17
	3.9 Role and Responsibility	17
	3.10 Location	17
	3.11 Commitment/Investment	17
	3.12 Focal Point	18
	3.13 Occupant Condition	18
	3.13.1 Gender	18
	3.13.2 Culture	18
	3.13.3 Age	18
	3.14 Other Factors	19
4	**Occupant Behavior Concepts: Cues, Decisions and Actions**	21
	4.1 Introduction	21
	4.2 The Protective Action Decision–Making Process	22
	4.2.1 Cues	23
	4.2.2 Sensing the Cue(s)	24
	4.2.3 Paying Attention to the Cue(s)	26
	4.2.4 Comprehending the Cue(s)	27
	4.2.5 Processing the Cue	28
	4.2.6 Decision–making and Taking Protective Action	31
	4.2.7 Breaks in the Decision-making Process: Seeking Additional Information	35

4.3		The Myth of Panic	36
4.4		Impact of Human Behavior in Fire on Fire Protection Engineering Design and Analysis	37
	4.4.1	The Impact of Human Behavior on Evacuation Timing Calculations	37
	4.4.2	Human Behavior Considerations Related to Warnings or Messages	38
	4.4.3	Human Behavior Considerations Related to Occupant Emergency Training	38
4.5		Summary: Behavioral Facts	38

5 Effects of Fire Effluent 41

5.1		Effects of Exposure to Smoke and Smoke Components	41
	5.1.1	Asphyxiants	42
	5.1.2	Hypoxia	43
	5.1.3	Carbon Dioxide	44
	5.1.4	Irritants	44
	5.1.5	Toxic Fire Gas Interactions	44
	5.1.6	Heat	45
5.2		Visibility/Smoke Obscuration	46

Part II Modelling Human Behavior in Fire

6 Development and Selection of Occupant Behavioral Scenarios 51

6.1		Introduction	51
6.2		Background	51
6.3		Occupant Behavioral Scenarios	52
	6.3.1	Some Aspects of Occupant Scenarios are tied to the Fire Scenario	52
	6.3.2	Identifying Occupant Scenarios	53
6.4		Documentation	54
6.5		Quantifying Occupant Behavioral Scenarios for the Evaluation	55
	6.5.1	Delay Time Before Occupants Begin to Evacuate	55
	6.5.2	Travel Speed	56
	6.5.3	Available Route Options	56
	6.5.4	Path Choice	56
	6.5.5	Travel Flow	56
6.6		Sensitivity Analysis	56

7 Calculation of Effects of Fire Effluent 57

7.1		Toxicity Analysis Methods	57
	7.1.1	Ct Product and Fractional Effective Dose	57
	7.1.2	Life Threat Hazard Analysis	66
	7.1.3	Typical Production Levels Based on Fire Type	67
	7.1.4	Susceptible Populations	68
7.2		Background and Guidance on Reduced Visibility Conditions	68

8 Physical Movement Concepts 73

8.1		Introduction	73
8.2		Factors That Impact Movement Time	74
8.3		Methods for Calculating Movement Time	75
	8.3.1	Hand Calculations	75
	8.3.2	Speed	75
	8.3.3	Specific Flow	77
	8.3.4	Total Flow Capacity	79
8.4		Examples	80

Contents xiii

| | | 8.4.1 | Example #1 | 80 |
| | | 8.4.2 | Example #2 | 81 |

9 Egress Model Selection 85
9.1 Introduction 85
9.2 Project Considerations 86
9.3 Model Attributes 87
9.4 Developing Model Scenarios 88
   9.4.1 Building Configuration (the "Structure") 89
   9.4.2 Population Configuration (the "People") 89
   9.4.3 Procedural Configuration 90
   9.4.4 Environmental Configuration 90
9.5 Model Output 92
9.6 Characterization of Current Computer Based Evacuation Models ... 93

10 Egress Model Testing 97
10.1 Introduction 97
   10.1.1 Relevant Work in Fire 98
10.2 Testing the Process 99
10.3 Pre-Model Execution 100
   10.3.1 Step A – Model Selection 100
   10.3.2 Step B – Model Configuration 102
10.4 Post-Model Execution 103
   10.4.1 Step C – Model Verification 104
   10.4.2 Step D – Model Validation and Calibration 106
   10.4.3 Sensitivity Analysis 110
10.5 Reporting Test Results 111

11 Estimation of Uncertainty and Safety Factors 115
11.1 Introduction 115
11.2 Sources of Uncertainty 115
11.3 Strategies for Managing Uncertainty 116
   11.3.1 Reduced Need for Safety Factors 116
   11.3.2 Increased Need for Safety Factors 117
11.4 Sensitivity Analysis to Reduce Uncertainty 117
11.5 Robustness to Reduce Uncertainty 118
   11.5.1 Evacuation Model Type 118
   11.5.2 Enclosure Representation 118
   11.5.3 Population Perspective 119
   11.5.4 Behavioral Perspective 119
   11.5.5 Model Validation 119
11.6 Considerations When Using Safety Factors 119

Part III Fire Situation Management

12 Enhancing Human Response to Emergency Notification and Messaging ... 123
12.1 Introduction 123
12.2 Human Response to Emergency Warning 124
   12.2.1 Processing Information 124
   12.2.2 Inhibiting Factors 124
12.3 Guidance on Emergency Communication Strategies 124
   12.3.1 Alerts 125
   12.3.2 Warnings 125
   12.3.3 Intelligibility 128
   12.3.4 Occupants that Remain in Place 129
12.4 Buildings with Limited Visual and Audible Notification Appliances ... 129

	12.5	The Use of Unannounced and Announced Emergency Drills	129
		12.5.1 Unannounced Drills	129
		12.5.2 Announced Drills	130
13	**Managing the Movement of Building Occupants**		**131**
	13.1	Introduction	131
		13.1.1 Persons Responsible for the Design of Buildings Before They Are Occupied	132
		13.1.2 Persons Responsible for the Operational Management of Occupant Movement After Buildings Are Occupied	132
	13.2	Available Resources for Tailoring Occupant Movement Strategies to Specific Buildings	132
	13.3	Factors and Assumptions Used to Divide Occupants into Groups that Require Different Movement Strategies	133
	13.4	Delayed Movement for Persons with Critical Functions	134
	13.5	The Model for Designing Buildings that Optimize Decisions About the Movement of Building Occupants	135
		13.5.1 Ways that Designer Can Use the Model	135
		13.5.2 Decision Process One: Which Groups Are Safe Where They Are Already Located?	137
		13.5.3 Decision Process Two: Where Are the Safer Locations?	137
		13.5.4 Decision Number Three: What Are the Means to Relocate Occupants to the Safer Location?	138
	13.6	The Decision Model for the Operational Planning of People Movement During Fire Emergencies	138
	13.7	Informational Inputs Common to Both Versions of the Model. The Number of Occupants in the Various Locations of the Building	140
		13.7.1 Locations of Building Occupants	140
		13.7.2 Projected Growth/Mitigation of Fire Hazards	140
		13.7.3 Building Features that Separate Stationary Occupants from Fire Hazards	140
		13.7.4 Building Features that Separate Moving Occupants from Fire Hazards	141
	13.8	Limitations of Occupants	142
	13.9	Procedural Assistance	143
	13.10	Using the Operational Model to Adapt the Plan Depending on How the Emergency Develops	143
		13.10.1 Examples of Circumstances that Can Disrupt Even Well-Conceived Plans	144
		13.10.2 Backup Strategies	144
		13.10.3 Using the Model During Fire Emergencies	145
		13.10.4 Definition of and Importance of Situation Awareness while Managing Occupant Movement During Fire Emergencies	145
	13.11	Guidance on Implementing the Models	146
		13.11.1 Informational Inputs Needs to be Acquired from a Variety of Sources	146
		13.11.2 A Checklist for Data Inputs to the Decisions in the Models	146

Addendum: Glossary of Terms 149

References 151

Introduction

The prediction of human behavior during a fire emergency is one of the most challenging areas of fire protection engineering. Yet, understanding and considering human factors is essential to designing effective evacuation systems, ensuring safety during a fire and related emergency events, and accurately reconstructing a fire.

In 1988, the Society of Fire Protection Engineers (SFPE) completed the First Edition of *The SFPE Handbook of Fire Protection Engineering*. The *Handbook* contained the initial building blocks for conceptual and analytical discussions directed to the fire engineering community on the treatment of human factors and behavior from the perspective of occupants in fire emergencies and building evacuation scenarios. The theory, data, studies and analytical approaches provided in four Chapters in 1988 presented the perspective and knowledge of the individual authors.

However, it was not until 2003 that a consensus-based document was published to provide guidance to the fire protection engineering community: the *SFPE Engineering Guide, Human Behavior in Fire* [1]. This Guide became a popular resource, providing a concise overview and summary of various topics that underlie how occupant characteristics and the social and fire environment influence the response, reactions, and movement of occupants in fire.

Human factors research and practical experience advanced substantially over the succeeding decade. SFPE has been monitoring this progress, and the SFPE Task Group on Human Behavior has made the effort to develop and advance an updated and expanded version of this Guide; the result is the 2018 Second Edition, *SFPE Guide to Human Behavior in Fire*. This Second Edition has been developed with the significant and extensive input of individuals with expertise in sociology, behavioral sciences, human factors, toxicology, egress model development, fire dynamics, and practical fire protection engineering. This revision builds on the topics in the First Edition and covers additional subjects, such as occupant behavioral scenario selection, toxicity assessment, visibility-in-smoke, emergency notification/messaging, management the movement of building occupants, model selection and egress model testing.

The goal of this expanded *SFPE Guide to Human Behavior in Fire* is to provide a common introduction to this field for the broad fire safety community: fire protection engineers, design professionals, and code authorities. The public benefits from consistent understanding of the factors that influence the responses and behaviors of people when threatened by fire and the application of reliable methodologies to evaluate and estimate human response in buildings and structures.

It is the further purpose of this Guide to lessen the uncertainties in the "people components" of fire safety and allow for more refined analyses with less reliance on arbitrary safety factors. As with fire science in general, our knowledge of human behavior in fires is growing, but is still characterized by uncertainties that are traceable to both limitations in the science and unfamiliarity by the user communities. The concepts for development of evacuation scenarios for performance-based designs and the technical methods to estimate evacuation response are reviewed with consideration to the limitations and uncertainty of the methods. This Guide identifies both quantitative and qualitative information that constitutes important considerations prior to developing safety factors, exercising engineering judgment, and using evacuation models in the practical design of buildings and evacuation procedures.

The technical advances that support this new edition of the Guide have been reported in numerous scientific and research publications, not necessarily written for and available to most individuals or organizations. The Task Group has distilled the most relevant and useful information into a consensus-based resource document whose key factors and considerations impact the response and behavior of occupants of a building during a fire event.

This 2018 Guide is designed as a singular resource for the fire safety community. The reader/user of this Guide is referred to the past editions and the current edition of the *SFPE Handbook of Fire Protection Engineering* [2] for additional readings,

© Society of Fire Protection Engineers 2019
SFPE Society of Fire Protection Engineers, *SFPE Guide to Human Behavior in Fire*, https://doi.org/10.1007/978-3-319-94697-9_1

more detailed discussions, and useful data that complement this Guide. The current Fifth Edition of the *SFPE Handbook of Fire Protection Engineering* contains 10 Chapters that are directly related to the content in this *SFPE Guide to Human Behavior in Fire*. Those Chapters to consider for further reading and the perspectives of individual authors are:

- Volume 2, Chapter 56, Egress Concepts and Design Approaches
- Volume 2, Chapter 57, Selection Scenarios for Deterministic Fire Safety Engineering Analysis: Life Safety for Occupants
- Volume 2, Chapter 58, Human Behavior in Fire
- Volume 2, Chapter 59, Employing the Hydraulic Model in Assessing Emergency Movement
- Volume 2, Chapter 60, Computer Evacuation Models for Buildings
- Volume 2, Chapter 61, Visibility and Human Behavior in Fire Smoke
- Volume 2, Chapter 62, Combustion Toxicity
- Volume 3, Chapter 63, Assessment to Hazards from Smoke, Toxic Gases and Heat
- Volume 3, Chapter 64, Engineering Data (i.e. Data sets for human behavior analyses)
- Volume 3, Chapter 76, Uncertainty

It is the intent of the SFPE Task Group on Human Behavior that the Second Edition of the *SFPE Guide to Human Behavior in Fire,* together with the Fifth Edition of the *SFPE Handbook of Fire Protection Engineering,* provide students, researchers, engineers, and public authorities with sound understanding of human behavior during emergencies. It is the belief of the Task Group that this understanding will facilitate the successful and cost-effective provision of life safety in emergency incidents.

Integrating Human Behavior Factors into Design

2

2.1 Use of this Guide

When considering human behavior related to fire, there are three key areas of interest as identified by Hall [312].

1. Behaviors that cause or prevent fires
2. Behaviors that affect fires; and
3. Behaviors that increase or reduce harm from fires

The focus of this SFPE Guide is primarily behaviors related to evacuation and refuge-finding or relocation to a safe area. These behaviors principally fall into the third listed category (behaviors that increase or reduce harm). However, human behaviors such as occupant firefighting or leaving doors open/closed that are associated with the second category can be critical and consequential to the evacuation/refuge process and should not be ignored.

Engineers are typically reliant on quantitative factors that result in deterministic designs and decisions. Often the assumptions and considerations related to human behavior and decisions during a fire are common generalizations or individual perceptions based on one's personal experiences. There certainly may be a degree of validity to such generalizations and perceptions; however, reliance on broad characterizations of human behavior may result in ignoring those factors and variations in the population (and among individuals) affected by a fire in a structure, building or other setting (e.g. wildland and wildland-urban interface fires). This SFPE guide intends to assist the engineer in avoiding easy generalizations and perceptions and, alternatively, guide the engineer to the quantitative factors and information needed for competent analysis of human behavior in fire events.

The fire protection engineer and others with an interest in fire safety (e.g. building owners, building/fire officials) may not have the specific and detailed background in those diverse areas of study—sociology, behavioral sciences, human factors, toxicology, egress model development—this factors into more comprehensive understanding and assessment of people's behaviors and decisions during a fire. This SFPE *Guide to Human Behavior in Fire* will serve to broaden the users' understanding of the factors that influence the responses and behaviors of people when threatened by fire. For further assistance, this Guide discusses and describes the application and limitations of recognized methodologies used to evaluate and estimate human response in fire scenarios. An important aspect to applying any of the methodologies in this guide is to develop an appreciation for the uncertainties involved and considerations related to the treatment of uncertainties. The use of information in this Guide will contribute to the fire protection engineer's confidence when confronted with investigations or assessments of human behavior in fire scenarios and building design. The Guide's content is applicable when prescriptive building designs (designs based on code assumptions) or performance-based designs are being implemented. Although building scenarios are a common context for the use of the concepts in this Guide, there are other scenarios and situations where the considerations and evaluation methods of the guide can be useful. Examples include tunnels, cruise ships, and wildland fire settings.

This book is organized in three major sections:

- Section I: Understanding Human Behavior in Fires—Chapters 3–5 presents general background and knowledge

© Society of Fire Protection Engineers 2019
SFPE Society of Fire Protection Engineers, *SFPE Guide to Human Behavior in Fire*, https://doi.org/10.1007/978-3-319-94697-9_2

- Section II: Modeling of Human Behavior in Fire—Chapters 6–11 provides guidance and discussion of quantitative approaches and methods for evaluating human behavior scenarios
- Section III: Fire Situation Management—Chapter 12 provides for an understanding of the use of emergency notification/messaging solutions and Chapter 13 presents guidance and considerations for managing occupant movement in emergencies

Within each section there are multiple topics. The user of this guide is advised to consider the chapters in order for a full and integrated understanding of the concepts and methods of analysis. The content and benefits of each chapter is summarized in Table 2.1.

Table 2.1 Beneficial Knowledge for User of this Guide

Guide chapter	Beneficial knowledge for user of the guide
Section I: Understanding Human Behavior in Fires	
Chapter 3: Population Characteristics	Understand the many types of characteristics that may influence occupant performance, reaction and movement process in an emergency situation: gender, age, culture, occupant density, role/responsibilities, physical/cognitive abilities, individuals alone or in groups, etc.
Chapter 4: Occupant Behavior Concepts: Cues, Decisions & Actions	Understand the theory and process of individuals or groups of occupants taking protective actions in the context of the building fire emergency timeline. Chapter 4 discusses the Protective Action Decision Model (PADM) and how it serves to inform our understanding of the flow of information to occupants via cues from the fire, building or people in a fire situation and the resultant decision-making processes that influence their protective action responses during the fire incident. The theoretical discussion is supported by case studies and experimental studies to reinforce concepts presented in Chap. 4
Chapter 5: Effects of Fire Effluent	Understand how fire effluents can lead to incapacitation of people subject to the hazards of a fire. This chapter provides an overview and explanation of acute physiological fire hazards affecting escape capability. The concepts address exposure to toxicants, irritants, heat and visibility loss in a fire and the potential impacts, including additive effects, on the human body at varying degrees of exposure.
Section II: Modeling Human Behavior in Fire	
Chapter 6: Development and Selection of Occupant Behavioral Scenarios	Obtain guidance to assist in developing occupant behavioral scenarios. This Chapter outlines a process to guide the engineer through the multiple considerations that may account for one or more defined occupant behavioral scenarios. The process involves identification of those factors that set physical (e.g. building) or operational conditions and define the environment in which occupants' behaviors will transpire, which includes an understanding and definition of the fire scenario. The Chapter provides guidance on defining occupant characteristics and group factors as a key step to establishing reasonable assumptions on the timing of occupant response to cues or alarms and the timing of subsequent actions or behavior prior to and during occupant movement.
Chapter 7: Calculation of Effects of Fire Effluent	Understand the technical basis and background regarding the application of a variety of well-known and recognized methodologies (equations) for evaluating the toxic impact of effluents from a fire. The toxic impacts from environments with CO, HCN, hypoxic (low oxygen), CO_2 and others are addressed. Methods to evaluate the impact of irritants and heat exposure are described in Chap. 7. Guidance for addressing reduced visibility due to smoke obscuration is detailed in the Chapter. Chapter 7 provides the engineer with worked examples of the various methods used for calculating tenability in fires (e.g. Fractional Effective Concentration (FEC), Fractional Effective Dose (FED))
Chapter 8: Physical Movement Concepts	Understand the basic concepts of occupant movement. This is important for estimating the time for occupants to move to a place of safety or refuge. Specifically, Chap. 8 addresses the period of time once purposive movement to safety begins. This Chapter also provides guidance and quantitative methods to address those factors that influence occupant movement—speed, occupant mobility, occupant density, travel path geometry, and boundary layer considerations. Examples of application of the quantitative relationships are provided that are informative to the user and illustrative of the concepts of occupant flow parameters and queuing that occurs during an evacuation.
Chapter 9: Egress Model Selection	Obtain understanding and guidance for the selection of one of the various methods of performing egress calculations. The modeling approaches addressed are movement models executed using hand calculations (see Chap. 8) or computational evacuation models that are complex and require use of tested computational software (see Chap. 10). The user of this Chapter will benefit by understanding the many considerations that affect the choice of a particular calculation approach or model, such as project data needed, model input data needed,

(continued)

Table 2.1 (continued)

Guide chapter	Beneficial knowledge for user of the guide
	complexity of scenario(s), detail of output needed, and need to assess the impact of behavioral assumptions. General characteristics for a wide variety of known models are presented.
Chapter 10: Egress Model Testing	Guidance is provided, for users of computational egress models, in the areas of selection, configuration, verification, and validation. Chapter 10 provides the user with the knowledge to determine if the model is capable of generating useful, appropriate, and credible results. The guidance provided is intended to aid the user in judging existing model testing and how to conduct their own testing in order to increase their confidence in the model's performance and the credibility of the results produced.
Chapter 11: Estimation of Uncertainty and Safety Factors	Develop an understanding of the various sources of uncertainty—measurement uncertainty, model-input uncertainty, intrinsic uncertainty, behavioral uncertainty. The user will benefit from a discussion on approaches to addressing uncertainty including minimizing the use of model components that pose uncertainty, obtaining a better understanding of the real-world phenomena, performing sensitivity analysis, and applying conservatism or safety factors. With proper considerations, the engineer should achieve more confidence in the results, be cognizant of addressing those factors that can vary slightly but result in large impacts, and yet be assured that various factors are not promulgated in such a manner to result in overly conservative results.
Section III: Fire Situation Management	
Chapter 12: Enhancing Human Response to Emergency Notification and Messaging	Understand how people process emergency warning information and examine factors that may inhibit the process. Guidance is provided for those having responsibilities related to emergency communications during a fire /emergency event. Chapter 12 discusses how to create and disseminate messages using basic communication modes including audible and/or visual technologies. The guidance in this Chapter can help engineers and systems designers improve their design of emergency notification and messaging systems, which can influence and reduce pre-evacuation and total occupant evacuation time.
Chapter 13: Managing the Movement of Building Occupants	Guidance and considerations for managing the movement of occupants is presented in the form of a model. The model presents a means to consider the numerous factors that play roles in the use of evacuation, relocation, and protect-in-place strategies when deciding how building occupants can best be safeguarded during fire emergencies. The model is represented as flow charts that show the decision processes that lead to recommendations about which occupants should remain where they are already located and which occupants should move to safer locations. Fire protection engineers, architects, building owners, and managers, can use this model to clarify the occupant movement strategies that the building is designed to support.

2.2 Human Behaviour Assumptions Within Fire Codes and Standards

Fire codes and standards provide a series of general requirements/recommendations for achieving a given level of fire safety for a variety of different types of buildings and occupancies. Typically, they are the primary means and source of reference by which fire safety is achieved or measured in the building design process. Within such documents fire safety is typically achieved through the specification and operation of the physical components (e.g. fire compartmentation) or systems (e.g. sprinklers, alarms etc.) within a building. Such specifications commonly assume or require a reader to assume people evacuating the building behave in a certain manner, typically predicated on the provision of building management procedures. For certain specific engineering applications these assumptions may be appropriate as they are relevant for the specific building design/occupancy given how people are likely to behave during an evacuation and the associated management procedures provided. Conversely there are other circumstances where these assumptions may be inappropriate due to the people evacuating being unlikely to behave in the assumed manner. The regulatory assumptions present a model of human behavior. To ensure fire design guidance is appropriately applied, fire engineers must understand the underlying associated human behavior assumptions in this model and whether these are appropriate to apply for a specific engineering application.

Human behavior assumptions are generally not explicitly stated within fire codes and standards; i.e. the underlying conceptual model is not always obvious. However, such documents may make reference to common types of human behavior in fire which should be considered and/or imply how people might behave during an evacuation. A reader of the codes and standards is therefore required to determine how people are assumed to behave for a given requirement and determine if the subsequent fire safety provision is sufficient for a project. The following lists describe potential assumptions regarding human behavior in fire which may be found in fire codes and standards. These may assume a reader has applied additional measures in the design process in order to fulfill the assumptions e.g. providing evacuation staff management requirements, providing evacuation training, provision of signage, etc. The list is not exhaustive and not necessarily always applicable: this will be dependent on the fire code being adopted and the building design where it is being applied.

- People will be able to perceive/comprehend a cue of the need to evacuate e.g. from a fire alarm signal, other people, fire/smoke, etc.
- People will be aware of the intended response to the fire alarm.
- People will begin to evacuate upon activation of the fire alarm in the local area.
- People will be physically capable of travelling the stated maximum travel distances for each occupancy during an evacuation.
- People will be aware/willing/able to use the available emergency exits.
- People will use the full width of the escape routes and exits available.
- People can and will proportionally distribute between the available escape routes and exits according to the throughput each provides.
- People will be aware of what refuge areas are, there location in the building, and be prepared to use them.

If a given human behavior in fire assumption within a fire code or standard is not appropriate for a specific engineering application then consideration must be given to:

- Determining the sensitivity of the building design to the human behavior assumption; and/or,
- Using an alternate form of analysis or a performance-based approach.

2.3   Performance-based Design and Human Behavior Considerations Worldwide

The *SFPE Engineering Guide to Performance-Based Fire Protection Analysis and Design of Buildings* [3] identifies an overall process for performance-based design. This *SFPE Guide to Human Behavior in Fire* is a complementary document that should be considered in several parts of the design and analysis process and may be applicable in the overall process. This Guide on human behavior will assist with the quantitative and qualitative analysis needed in the performance-based design process where the protection of human lives is a fire safety objective.

SFPE's general process for performance-based design includes the following steps:

1. Define Project Scope
2. Identify Goals
3. Define Objectives
4. Develop Performance Criteria
5. Develop Fire Scenarios and Design Fire Scenarios
6. Develop Trial Designs
7. Prepare Fire Protection Engineering Design Brief
8. Evaluate Trial Designs
9. Select the Final Design
10. Design Documentation
11. Manage Changes

During the defining of the project scope, the general conditions of the situation/scenario should be determined.

- Building context (e.g. area, height, occupancy, hazards, construction, systems, egress features, etc.)
- People of interest (e.g. those exposed to effluent, those remote from fire, responsible staff, emergency responders, etc.)
- Occupant characteristics of those "people of interest"
- Building emergency response strategy
- Fire scenarios of interest

Table 2.2 International Selection of Engineering Guidance Documents Addressing Fire-Related Human Behavior

United Kingdom	**PD 7974-6** 'The application of fire safety engineering principles to fire safety design of buildings. Human factors. Life safety strategies. Occupant evacuation, behaviour and condition (Sub-system 6)' and CIBSE, *Guide E: Fire Safety Engineering*, "Chapter 7 - Means of Escape and Human Factors", Chartered Institution of Building Services Engineers 2010
New Zealand	**C/VM2** *Verification Method: Framework for Fire Safety Design*, For New Zealand Building Code Clauses C1-C6 Protection from Fire; New Zealand Ministry of Business, Innovation and Employment, 2013
Australia	"International Fire Engineering Guidelines", Edition 2005, Chapter 1.8 Occupant Evacuation and Control, Australian Government, State and Territories of Australia, 2005 (Collaboration of National Research Council of Canada, International Code Council (USA), Department of Building and Housing-New Zealand, Australian Building Codes Board)
Japan	"Comprehensive Fireproof Building Design Methods, Volume 1", Architectural Center of Japan, 1989
Nordic Countries	**INSTA/prTS 950**, Technical Specification 950 "*Fire Safety Engineering – Verification of fire safety design in buildings*" Horingsfrist, 2013
ISO	**ISO/TR 16738:2009** Fire-safety engineering -- Technical information on methods for evaluating behaviour and movement of people, International Organization for Standardization, Geneva, Switzerland, 2009
ISO	**ISO 13571:2012** Life-threatening components of fire -- Guidelines for the estimation of time to compromised tenability in fires, International Organization for Standardization, Geneva, Switzerland, 2002
ISO	**ISO/TS 29761:2015** Fire safety engineering -- Selection of design occupant Behavioral scenarios, International Organization for Standardization, Geneva, Switzerland, 2015

When defining the project scope, a determination will be made as to whether or not protection of people will be contemplated by the design. Designs where the goals only relate to property protection, mission continuity or environmental protection, such as might be the case in an unmanned storage or warehouse facility may typically not require the types of analyses identified in this guide.

With the development of performance-based analysis and design methodologies and guides during the last 30 years, there has been clear recognition of the need to address human behavior in fire protection engineering. From a worldwide perspective, there now exists guidance and methodologies for considering human behavior that had been lacking in the 1980s. Organizations from Asia, Australasia, Europe, and North America have contributed to this increased attention to human behavior. Table 2.2 lists a number of the notable international engineering guidance documents that provide designers direction, data and/or design performance objectives to address human behavior during fire.

The common fire safety objective related to human life as addressed by these documents is simply that a building design reasonably affords sufficient time to evacuate or reach a place of safety before encountering fire conditions that would result in serious injury or death. One important distinction that is often noted is that occupants intimate with the source of fire or burning materials are effectively excluded from consideration. Consequently, the scenarios of human escape addressed by this and other guides are those scenarios where the occupant is not intimate to the fire.

The guidance presented in this SFPE Guide, other performance codes and international fire documents considers six key topics that influence the analysis of human behavior.

- Building Characteristics
- Evacuation Strategies and Procedures
- Occupant Characteristics
- Occupant Behavior/Actions Preceding Evacuation
- Occupant Behavior/Actions During Movement/Relocation
- Occupant Exposure to the Fire Environment

Both the building characteristics and evacuation strategies are relatively static or preplanned features that should be identified and are important to anticipating and assessing occupant behaviors that may influence evacuation time. Given a clear understanding of the building and evacuation strategy (or lack of) allows for occupant characteristics, occupant behaviors/actions, occupant movement and occupant fire exposure to then be collectively evaluated. It is important to recognize that there is a degree of complexity, due to the inter-relationship of these factors, which can make it difficult to accurately and precisely predict the behavior of each and every individual.

While much of the material in this Guide is calculation-oriented, it is important to understand that integrating human behavior into the design process is not simply a matter of engineering calculations with demonstrable outcomes. To address the fire safety of occupants in a building, it is important to understand and consider the factors that may influence the responses and behaviors of people in threatening fires. The anticipation of human behavior and prediction of human responses during a fire event is one of the more complex areas of fire protection engineering.

Use of this guide will provide understanding of the factors that influence the responses and behaviors of people when threatened by fire and the application of reliable methodologies to evaluate and estimate human response in fire scenarios. With this information, it is possible for engineers to give better treatment and consideration in design than has previously been the case. Often, engineers and designers have made basic assumptions about the population or occupant behavior that may have little or no basis in behavioral literature. An example is the frequently cited assumption of occupants automatically and immediately evacuating a building upon the sounding of the fire alarm system. While the assumption may be appropriate for some occupancies having a practiced and leadership-oriented emergency plan (e.g., schools), such an assumption has been shown to be erroneous for occupancies that lack such an organizational structure [4, 5].

This guide can be useful but may not be sufficient to address those buildings that are occupied by populations that are provided with extensive training, require specific emergency response tasks or have dictated evacuation policies. These situations can be found in correctional facilities, hospitals, industrial plants, external storage yards, power stations and specialty laboratory facilities (e.g., nuclear/radiological, biological, explosive, etc.). For these type of situations, many portions of the Guide may be utilized, but users should do so with a full understanding of the specific fire scenarios, populations, training, response activities and evacuation approaches employed.

2.4　Time as Function of Behavior

Time is the basic engineering measure of evacuation or refuge finding; the fire protection engineering community has focused on time as the key issue to be addressed when considering human behavior. A basic representation of the fire emergency timeline that depicts the critical human responses or behaviors that impact and contribute to the evacuation process is illustrated below in Fig. 2.1. This Guide will assist in understanding potential factors that impact the pre-evacuation period, and how to translate the real-world considerations for the pre-evacuation period and the movement period into an engineering analysis.

Figure 2.1 is the basic timeline used in Chap. 4 of this Guide to explain the theory of human behavior in fire. In terms of behavior, it is inherent in the timeline that mental processes and actions are a continuum of information processing and decision making. This is an important concept as during any evacuation, occupants are subject to receiving, recognizing and interpreting cues that may impact their decisions before and during their movement towards a safe location. Once travel or movement begins, other additional cues and decisions may be relevant. For example, during the movement period other cues (e.g. smoke, occupant communications) and exposure to heat/smoke/toxic gases may be encountered that may influence occupant responses and in turn movement time; these factors should be considered to the extent possible. Chapter 4 of this Guide, "Occupant Behavior Concepts – Cues, Decisions & Actions" details the psychological and sociological processes that can take place during the entire evacuation timeline.

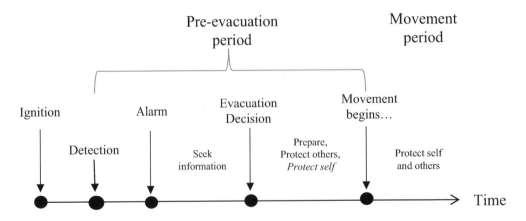

Fig. 2.1 Building Fire Emergency Timeline, that Displays an Example Building

2.4 Time as Function of Behavior

The process of evaluating the crisis, deciding on actions (or inaction), and taking those actions can be repeated multiple times between the Alarm time and the end of the Movement stage. As new information becomes available, or as prior information becomes interpreted differently, evacuees might well change their behavior. For example, the 619-people remaining above the aircraft impact floors of the South Tower of the World Trade Center might have believed that all hope of downward movement had been cut off. However, some discovered a path through an only partially damaged stairwell. Eighteen people took sufficiently prompt action that they went down 80 or more floors and escaped the building in the 56 min before the building collapsed [6].

The building fire emergency timeline generally depicts the evacuation process wherein occupants become aware of a building fire emergency and experience a variety of cognitive and social processes/actions before and while they travel to reach a place of safety. During the pre-evacuation and movement periods, occupants may be exposed to fire gases and smoke. The potential degradation in human response and incapacitation resulting from exposure to toxic gases, irritants, heat exposure and/or loss of visibility is explained in Chap. 5 of the Guide. While occupant exposure to fire effluent may degrade occupant response, it can also reinforce the sense of urgency to take appropriate action. The end goal represented by the timeline development is to determine if individuals can or cannot arrive at a place of safety, which is embodied in the concepts of Available Safe Egress Time (ASET) and Required Safe Egress Time (RSET) defined as follows [7].

Available Safe Egress Time (ASET) – The calculated time interval between the time of ignition and the time at which conditions become such that the occupant is estimated to be incapacitated, i.e. unable to take effective action to escape to a safe refuge or place of safety (time available for an individual occupant to escape or move to a safe location).

Required Safe Egress Time (RSET) – The calculated time interval required for an individual occupant to travel from their location at the time of ignition to a safe refuge or place of safety (time required for an individual occupant to escape or move to a safe location). Predicting RSET typically involves estimating the time that it would take for people to be notified that there might be a fire, the time that people would take for pre-evacuation activities such as alerting others, checking on others (e.g. family members), etc. and the time it would take for people to egress to a safe location. During the course of the evacuation, there may be other behaviors, actions or inactions that extend or reduce the time for escape.

The user of this Guide should recognize that the definitions of ASET and RSET focus on the "individual" as each occupant in a fire scenario will have their own ASET and RSET value depending on the individual's characteristics (e.g. age, physical capabilities), location at the time of a fire and path of travel relative to potential exposure to fire effluent. The ASET value is to be compared against RSET with the goal that (ASET) be greater than (RSET). By estimating ASET minus RSET for each person, it is possible to estimate how long it would take for the last person to reach safety. This approach can take extensive calculation time, since ASET minus RSET must be calculated for all individuals who are at above average risk. A limitation of this approach is that it may be difficult to estimate the rate of escape progress for the least capable evacuees.

In many cases, a simplification is used in which the occupants of a space (e.g., an auditorium, a suite of offices, or even the entire building) are treated as a uniform group. ASET and RSET for the group are estimated using a single set of values for human incapacitation and a single set of assumptions regarding occupant characteristics. In practice, the occupant assumptions and the values for incapacitation are generally intended to be conservative. This approach reflects the lack of foreknowledge of the types of people who might be in the building at the time of an emergency. Depending on the degree of conservatism applied, the values and assumptions selected should represent the mean or some value greater than the mean. In this case the value of ASET minus RSET is the time interval in which half or more of the group would reach safety. By using an equation for the distribution of the values of these characteristics, often based on a normal distribution, it is possible to adjust the estimated value of ASET and/or RSET to enable a large fraction of the group to reach safety. Its limitations are that (1) the actual distributions of the occupant characteristics are unknown and (2) a normal distribution does not go to a value of zero, so the value of RSET minus ASET for which 100% safe egress occurs may be extremely long.

To more fully address the considerations for occupant behavior during a fire, this Guide adopts an approach that requires the integration of qualitative and quantitative data and engineering methods to evaluate the time for evacuation or movement to a place of safety.

Five basic tasks are generally needed:

1. Define building characteristics and building evacuation strategy.
2. Review occupant characteristics and identify occupant groups' assumptions.
3. Develop pre-evacuation time assumptions.
4. Evaluate/calculate movement time.
5. Evaluate/calculate the fire environment impact on exposed occupants.

While modeling approaches continue to develop, there currently exists important qualitative and quantitative information that can facilitate and guide the analysis and estimation of occupant evacuation or escape. Qualitative information and theoretical background in Chaps. 3, 4, 5 and 6 is useful in defining the parameters of an evacuation scenario, identifying critical human factors, developing assumptions for model inputs and supporting engineering judgments. In some cases, the research or data found in the literature may fit the specific context of an evacuation scenario and provide highly relevant quantitative data applicable to the design situation. In other cases, the occupant characteristics and scenario may not fit with any available data which may require the analysis of more variables and scenarios to justify a result (e.g. RSET/ASET).

Currently, there are a number of tools and competent approaches available to the engineering community in the form of occupant movement calculations, toxicity analysis, software-based hydraulic models; and, complex behavioral occupant movement models which combine one or more aspects of occupant behavior and response. Chapters 7, 8, 9, 10 and 11 of the Guide provide background on the quantitative aspects of these approaches/models and important information related to the selection, validation and verification of models. This guide provides references to many sources for both qualitative and quantitative information needed by the fire protection engineer; however, additional detailed data and references are to be found in the SFPE *Handbook of Fire Protection Engineering* [2].

The management of the fire situation is an important consideration for human behavior analysis. The presence or lack of emergency warning or messaging systems (e.g. horns, bells, strobes, voice messages, and text messages) can impact the assumptions regarding the effectiveness and timing of occupant response. In Chap. 12, the background for understanding human response to warning systems and the steps that lead to an occupant's decisions to take protective action is presented. These factors should be accounted for when defining assumptions or impacts on the pre-evacuation period as influenced by the type and effectiveness of an existing or proposed warning system. The guidance in Chap. 12 provides information about the capabilities, limitations, and use of the various forms of alerts and warnings; this is valuable when planning for a new building or performing an analysis of an existing building. The factors and understanding of notification concepts provided for the engineer should assist the engineer in implementing and evaluating the impact of alerting and warning systems used in the built environment.

At the time of an emergency fire event, occupants may act on their own or be influenced by the training and evacuation exercises to which they have been exposed. For buildings with significant occupant loads (e.g. tall buildings) or vulnerable populations (e.g. hospitals), the building design and fire safety systems will play a significant role; these scenarios generally require the development of a pre-planned strategy. Such strategy and emergency plans should be relevant to the design/

Table 2.3 Factors and Considerations in Human Behavior Analysis

Factors and considerations in human behavior analysis* (*list is representative but not all inclusive)	
Building characteristics	Evacuation strategy/procedures
• Building type and use • Physical dimensions • Geometry of enclosures • Number and arrangement of Means of Egress • Architectural characteristics/complexity • Lighting and signage • Emergency information systems • Fire protection systems	• Total, zoned or staged evacuation • All or few occupants trained or drilled in procedures • Provisions for those with access and functional needs – infirm, disabled, incarcerated • Frequency of training or drills • Who is trained or drilled • Defend-in-place • Relocation
Population characteristics (See Chap. 3)	Fire Environment (See Chaps. 5 and 7)
• Population and density • Individuals alone or groups • Familiarity with building • Distribution and activities • Alertness • Physical/cognitive abilities • Role/responsibilities • Location • Commitment to task • Focal point • Gender • Culture • Age • Prior fire/evacuation experience	• Smoke and toxic gases • Temperature • Visibility • Transport, exposure, duration

features of the building and give consideration to the actions of emergency responders. Chapter 13 reviews the numerous factors that play roles in the use of evacuation, relocation, and protect-in-place strategies when deciding how to best safeguard building occupants during fire emergencies. Chapter 13 addresses the decision processes that lead to recommendations about which occupants should remain where they are already located, and which occupants should move to safer locations. The information in Chap. 13 is directed to and will benefit both building planners/designers and building operations staff and emergency responders who plan for and manage the operational movement of occupants during fires.

2.5 Informing the Engineer for Improved Consideration of Human Behavior

In past decades, the practicing fire protection engineer has not often encountered concerns for detailed and objective considerations of human behavior in building design. Usually serious consideration of human behavior issues was relegated to behavioral scientists and human factors researchers. Today, with the promotion of performance-based design, the importance of human factors and behavior in building design has entered the mainstream of fire protection engineering practice. This is evidenced by the many international documents that now address human behavior related to evacuation. To further the integration of human behavior into engineering practice, this SFPE Guide is dedicated to the topic of human behavior in fire and the many factors and considerations (see Table 2.3) that may play a role in such analysis.

Practicing engineers should avail themselves of the many sources of information on human behavior. There is a large resource of information available to address human behavior issues related to building design and fire safety systems applications. It is hoped that the information in the following Chapters of the Guide will well serve the fire protection engineering professional by providing guidance on the quantitative and qualitative human behavior analysis needed in the performance-based design process.

Part I

Understanding Human Behavior in Fires

Population Characteristics

3

3.1 Introduction

This chapter identifies factors necessary for considering human behavior in fire. These factors include occupant characteristics such as gender, age, physical capabilities, sensory capabilities, familiarity with the building, past experience and knowledge of fire emergencies, social and cultural roles, presence of others, and commitment to activities.

The occupant characteristics of a building population need to be reviewed to identify the occupant group or groups that are important to consider in the analysis. Using the list of occupant characteristics, a group or groups can be distinguished by attributing key characteristics. Not all characteristics are essential factors, but those that are critical and expected to influence the reaction and behavior of a group or groups should be noted.

In performing the evacuation analysis, it may be possible to rely on a single defined occupant group that is recognized as the most critical and is conservatively characterized. However, it may also be necessary to perform additional analysis when several identified occupant groups, in a given building, are distinguished by their varying characteristics.

3.2 Population Numbers and Density

The occupant load of a room is the maximum number of persons anticipated to be present for a given configuration or use. This may be defined according to design requirements for the room (specified by design specifications), number of seats, or by dividing the area by an appropriate occupant load factor. These occupant load factors are generally established in adopted codes and regulations. Where actual occupancy load data are available for similar occupancies, these may be used. Potential changes in occupancy or use need to be considered. Conservative design requires use of the maximum potential occupant load. It is important to be aware that there can be a large variation and uncertainty surrounding estimates for size and density of the occupant population,

Situations exist where codified information is not sufficiently accurate for the particular design under consideration [8]. In such circumstances, the designers should access other data sources or generate the data by carrying out surveys of similar premises. Designers should be mindful that the numbers and distribution of occupants in a building will change with time and activity. In a hotel, during the night every sleeping room in the hotel might be occupied while during the day, the sleeping rooms may be occupied by only a handful of housekeeping staff as they do their day's work. This occurred in the Vista Hotel during the bombing and fire at the World Trade Center in 1993 [9, 10]. Similarly, an arena may have its stands filled during a performance but have its stands empty and its floor space filled during an exhibition.

The number of people using a building or space and their distribution or density in that space will affect travel speeds (See Chap. 8).

© Society of Fire Protection Engineers 2019
SFPE Society of Fire Protection Engineers, *SFPE Guide to Human Behavior in Fire*, https://doi.org/10.1007/978-3-319-94697-9_3

3.3 Alone Or with Others

The presence of other people will influence behavior and decision making. Response to alarms or fire cues is affected by whether people are alone or with others. The presence of other people can have an inhibiting effect on the definition and initiation of action from initial ambiguous cues [11]. However, the presence of other people may increase the chances of a person being notified of an emergency and allow for group decisions of what actions to take.

There is a phenomenon referred to as 'social influence' that plays a role here [12]. There are two types of social influence – normative and informational. With normative social influence, people try to meet social norms and do what they think is 'expected' of them. This can result in people ignoring alarms when others do not act, in order to avoid standing out or looking foolish. People who are alone tend to respond more rapidly to ambiguous cues [13]. Informative social influence is where people observe others in order to better understand a situation and possibly follow their behavior. This has been shown, for example, to encourage herding behavior in exit choice [14, 15].

3.4 Familiarity with the Building

Occupant response may be influenced by familiarity with a building and its systems. In some conditions, frequent users of a building may have a complete knowledge of the nearest and alternative egress routes and warning systems. They may be expected to make an effective evacuation, particularly if subjected to regular emergency training and evacuation drills. Infrequent users of a building, such as members of the public, will usually depend more upon signs and staff. These users may be less familiar with, and less responsive to, warning systems. Members of the public may also be more likely to attempt to leave by the route they entered the building once they have decided to evacuate, unless they are directed otherwise by signs or systems.

Providing exit signs to indicate egress routes does not ensure that occupants will notice the signs or will use these exits to leave the building [16–18]. Since people are continually exposed to exit signs but are rarely, if ever, required to use non-familiar exits, they may have learned to filter out this information during normal usage at it provides little functional use that impacts everyday decision making. During an emergency, it is unlikely that occupants will be prepared to try a route they have never used before to leave the building. Occupants are more likely to attempt to leave by their familiar route for which they know the location, conditions, and location of discharge. If their familiar route out of the building is judged impassable due to conditions such as smoke or crowd, then occupants might rely on exit signs to find an alternative way out. In such conditions, the exit signs will need to be highly visible and conspicuous to be distinguished from surrounding information and be easily noticed by occupants.

Generally, it cannot be expected that occupants of buildings are familiar with all the emergency exits of the building, unless they have been trained as to their location and use.

3.5 Distribution and Activities

The evolution of an evacuation event will depend upon the extent to which occupants are evenly distributed throughout the occupied spaces or concentrated in particular locations. The initial response may be affected, to some extent, by the activities the occupant is engaged in immediately before the fire. It is important to obtain pre-evacuation time data for occupants engaged in different activities (such as eating in a restaurant, shopping, watching a film or entertainment, sleeping, or working) [4]. The activities that people are engaged in affect their response and how long before they take to disengage from the task before taking protective action. (See Sects. 3.11 and 3.12 in this chapter on commitment and focal point). People who are sleeping or showering, for example, when they are notified of an emergency will need additional time to waken and dress for the outdoors.

The distribution of occupants throughout a space will impact movement speeds (the more people, the denser the movement flow and the slower the walking speeds). Occupant density can also impact the ability to communicate instructions. Highly dispersed occupancies may create difficulties in communication; however, densely occupied or noisy occupancies may also hinder the ability to communicate.

3.6 Alertness

When people are in bed or asleep, their response times to a fire alarm can be expected to be considerably delayed. Drug or alcohol use also affects alertness. Additionally, the involvement of people in activities being carried out within the building or their interaction with the other occupants of the building can affect their awareness of other circumstances [19, 20].

3.7 Physical and Cognitive Ability

In many buildings, a proportion of the population may be impaired (cognitively and/or physically) or will present some level of limitation related to injury, illness, poor health, or other medical conditions. Occupants' preexisting medical conditions may influence tolerance to fire effluents. Some of the occupants may have to rely entirely on assistance of others or may not be able to be moved [21–24].

The initial response of disabled people may involve a considerable preparation time prior to movement. The movement of disabled occupants is significantly influenced by the nature of their disability and building elements such as doors, ramps, and stairs. People with a hearing disability may require special means of notification of a fire, although their evacuation movement may not be different than mobile occupants. People with a visual disability may perceive audible information such as a fire alarm or a voice communication message but might need assistance to find a suitable evacuation route. However, once in a stairwell, they might move independently at the speed of the group.

3.8 Social Affiliation

Social affiliation refers to the degree to which people feel connected to others and can influence a person's desire to warn and protect others or to seek to gather with others with whom they have social or emotional ties (e.g., family, friends, and colleagues). Separated group members are likely, first, to attempt to re-establish the group before moving towards the nearest exit [25]. In addition, the speed of movement will often be dictated by that of the slowest member of the group.

Groups have been found to affect the smooth merging of flows in corridors as they attempt to remain within close proximity. For example, rather than each individual in the group merging singly into the flow, the group will try to stay together [26].

3.9 Role and Responsibility

The roles and responsibilities of occupants during the normal use of the building will, in an emergency, influence their behavior and the behavior of others. Sufficient, well-trained, and authoritative staff will shorten the ambiguous, information-gathering phase of pre-evacuation time.

3.10 Location

Individual occupant responses are influenced by their specific location in relation to the fire, the warning system, and the escape routes. Location can influence the time to notification, comprehension of the alert signal or message, and actual travel distance.

3.11 Commitment/Investment

People are action- and goal-oriented. They have reasons for being in a particular place. Those reasons may continue to guide their behavior even when an emergency occurs. People who have invested a significant amount of time money, or effort to their activity (for example, waiting in a queue for service or waiting for a meal in a restaurant) will be more reluctant to disengage that task in response to an alarm signal if that means that they must lose their place in line or walk away from a meal [25, 27].

3.12 Focal Point

If the setting has a particular focal point, such as a stage in a theater, the population of the building will normally look to that point for guidance in the first moments of an alarm and evacuation [28].

The focal point can be used positively in an emergency, recognizing that people will often look to the stage, lecturer, etc. for "permission" to leave or indication of appropriate exit routes to use.

3.13 Occupant Condition

Throughout the evolution of the fire, an occupant's ability to respond will depend on their cumulative and instantaneous exposure to fire effluent.

3.13.1 Gender

In general, females are more likely to alert or warn others and evacuate in response to fire cues than males [29, 30]. Similarly, in the residential environment, it has been found that men were more likely to fight the fire, while women were more likely to gather family members and call the fire department. [30, 31]. A study of occupant behavior in health care facilities also found a higher likelihood of males attempting to fight the fire while female staff were more likely to take protective actions and rescue patients, consistent with their training and assigned responsibilities [32]. The re-entry behavior that was observed among men in residential fires was noted to frequently agree with the social role of the male as protector of the family [33].

Some of these studies were undertaken more than 40 years ago, and to some degree it may be that the findings were influenced by a more rigid assignment of roles by gender than exists today. For example, the Project People II final report notes that the gender distribution of the participants in the health-care study reflected the distribution of occupations among the participants - the study group was largely made up of women, and the predominant occupation among the participants was nursing. The protective actions taken by the women in the study may have been a reflection of their role as nurses (and caregivers) rather than their gender. Today, more and more nurses are men, many women work as security officers and medical doctors, and a health-care study done today might show that the protective actions are taken by nursing staff while the firefighting roles are taken by security staff, and gender may be less relevant. Similarly, in residential fires, the protective actions that have long been assigned to women may actually be shared more equally in households where parental responsibilities are shared.

The above discussion, however, serves only as a reminder that the effects of gender and the effects of role can be muddied in some older studies and users should be aware of that. Further studies should be done to help differentiate the influence of gender and the influence of role as society changes, and gender-assigned roles are less prevalent.

3.13.2 Culture

There is limited research available on the influence of culture on evacuation performance (one such resource is provided by Ozkaya) [34]. However, culture can be assumed to influence factors such as social affiliation, role, and responsibility, given research from the field of organizational behavior [35–37].

3.13.3 Age

Variations in evacuation performance as a function of the age of the individual can be expected and should be accounted for. With the aging population trend, this factor bears stronger consideration, especially given that data for developing evacuation times typically used college age students. Kose [38] categorizes the expected performance differences using three categories; sensory skills, decision making, and action (e.g., mobility, swiftness, etc.). These three categories can be expected to affect all evacuation phases, from recognition through actual movement throughout the structure and decision-making along the way. Furthermore, this influence can be expected not to be monotonic, but rather with evacuation performance degradation for the very young and the very old. Unfortunately, while the effect of limitations in each of the three categories on evacuation performance can be somewhat readily quantified (for example, in factors such as physical and cognitive ability), the linkage between age and these degradations are less established.

Age can also be expected to influence walking speed and the ability of the individual to withstand exposure to by-products of fires. Specifically, the very young and the elderly in poor health may be less able to withstand the debilitating effects of smoke and heat [39–41].

3.14 Other Factors

There are other factors that influence people's behavior that are more environmental than occupant-based that also should be considered. These include external factors such as weather, which can inhibit evacuation (e.g., reluctance to exit a building during a thunderstorm or heavy rain, causing a bottleneck at the exit door) and/or increase pre-evacuation delay time (e.g., additional time to dress for cold winter weather). Other factors to consider are occupant reaction to uneven floor or wall surfaces, complex evacuation routes, lighting levels and noise levels of alarms or use of strobe lights in corridors.

Occupant Behavior Concepts: Cues, Decisions and Actions

4

4.1 Introduction

The purpose of this chapter is to present the theory of human behavior in fires, which includes the occupant decision-making process and the resulting beneficial or potentially harmful actions. Protective actions can consist of any action or behavior performed during the building fire evacuation timeline that assists the individual or group of individuals to protect him/herself and/or others from the dangers of the fire event.

This chapter provides guidance on understanding the actions that occupants take during the building fire emergency timeline (shown in Fig. 4.1).[1] The timeline begins with ignition of the fire, encompasses the pre-evacuation period,[2] and ends with the movement period. The pre-evacuation period is the time between receipt of fire cues by a member (or members) of the building population[3] and a decision to take protective action. The pre-evacuation time also includes the time during which protective actions are performed, including preparatory actions (e.g., gathering personal items, getting dressed), actions to protect others (e.g., warning others, assisting), and actions to protect oneself. Actions to protect oneself are particularly relevant in cases where evacuation is not possible. Here, a building occupant may elect to *protect him/herself* during the pre-evacuation period by "sheltering-in-place," which could involve placing wet towels around the door to prevent smoke spread. If movement to safety is possible, the movement period also consists of actions to protect oneself (i.e., movement to a place of safety, either outside of the building or some other location of safety within the building), and/or actions to guide or assist others. Although the timeline is shown in a linear fashion, it should be noted that the evacuation (or protective action) decision-making process can be circular in nature, in that the receipt of new information from the fire event can alter or confirm original decisions made (described further in Sect. 4.2).

Depending upon the circumstances, for example, toxic conditions or the actions of others, occupants' decisions and/or actions can take a considerably long time to complete and will contribute to the time to reach safety. The information provided in this chapter can help users/readers to more accurately account for these actions, including which actions may take longer than others and why, when calculating evacuation timing in a proposed design building fire. In addition, users who are working to create evacuation or safety procedures, including emergency communication procedures or messages, will also find this chapter useful in understanding the aspects of emergency information that positively (or negatively) affect decision-making and response, and in turn, evacuation time.

This chapter will begin with a discussion of the process by which occupants in building fires make decisions and take actions in emergency conditions. Case studies will be provided throughout the chapter to provide examples of the theory and ideas being presented. Next, the myth of panic will be discussed, followed by the impact of human behavior in fire on engineering analysis and design. This section on impact will detail the ways in which the information presented in this chapter

---

[1] This schematic displays a scenario of a fire within a building where an alarm system is present and the safest option for building occupants is to evacuate the building. Other scenarios, for example, whereby a different type of alerting/warning system is used and/or another safety goal is to be achieved, would require different labels for the timeline shown in Fig. 4.1.

[2] Other terms have been used to express the pre-evacuation period, including pre-movement or pre-response

[3] The receipt of fire cues by staff, for example, is labeled in Fig. 4.1 as "detection," and the receipt of fire cues by the larger population is labeled in Fig. 4.1 as "alarm."

© Society of Fire Protection Engineers 2019
SFPE Society of Fire Protection Engineers, *SFPE Guide to Human Behavior in Fire*, https://doi.org/10.1007/978-3-319-94697-9_4

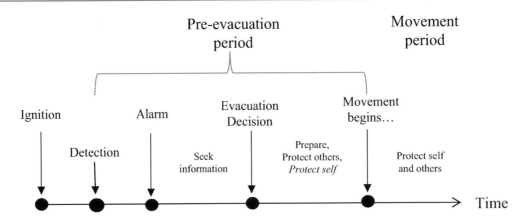

Fig. 4.1 Building fire emergency timeline, that displays an example building notification type (i.e., fire alarm)

will better inform engineering practice – identifying why the engineer should account for human behavior in fire in their projects. A final section provides a summary of the major findings captured by this chapter, referred to as the "behavioral" facts of human behavior in fire.

4.2 The Protective Action Decision-Making Process

Frequently, when considering human behavior in fire, there is an assumption made that building occupants respond immediately to initial cues of a fire. This assumption would be incorrect, in many cases, and would provide inaccurate results from egress calculations. Instead, there is considerable evidence in the literature to the contrary [4, 42, 43, 44], and taking that evidence into account when assessing a building's design would likely improve the life safety afforded its building occupants.

The Protective Action Decision Model (PADM), which is based on over 50 years of empirical studies of hazards and disasters [45, 46, 47, 48], provides a framework that describes the information flow and decision-making that influences an individual to take protective actions in response to natural and technological disasters [49]. The process can occur individually or, more likely, within a group (otherwise known as the milling process). The PADM has been adapted here for use in building fires.

In this model, people take time sensing cue(s) from the physical and social environment, paying attention to the cue(s), attempting to comprehend the cue(s), and then processing the threat and risk to make a decision on what to do next. This is referred to in this chapter as the process of taking protective action in a building fire. Depending upon how the threat is processed (e.g., what is going on and how dangerous is it?), an occupant will either a) seek additional information, b) engage in actions to protect people or property, or c) resume normal activities; all three shown in boxes in Fig. 4.2. Each time an individual receives new information from his/her physical/social environment, the decision-making process can begin again, either to alter or confirm original decisions made, if any. Note – that there are characteristics of the building occupant that will likely affect how they participate in this decision-making process, known as "receiver characteristics." For example, if a person is hearing impaired, this may influence how they receive the information that is provided to them in a building fire. If a person has physical disabilities, that may affect his/her risk perception, decisions, and/or actions performed. Additional receiver characteristics to keep in mind are the following: physical or cognitive disabilities, their previous knowledge and experience in fire events or other types of emergency, the types of actions in which they are engaged at the time of the fire, and relationships with others in the building.

The PADM, which has been adapted to building fires, is redrawn in Fig. 4.2. A more in-depth description of the PADM, and the additional research and social models that it draws upon, such as the Mileti and Sorensen warning model, are described in other references [50].

Figure 4.2 shows an idealized process of decision-making in a building fire. The text included in this chapter discusses how this decision-making process can be compromised by the decision-maker (i.e., the building occupant) and the physical and social environment surrounding the occupant.

The following sections will describe the stages of the decision-making process. One section of this chapter will be devoted to each stage. First, describing the stage in more detail. Second, the influence of the three elements of a building fire event: the fire environment, the building, and the people, on the decision-making stage will be discussed. The first stage of the decision-making process is the presence of cues (or information) from the fire event.

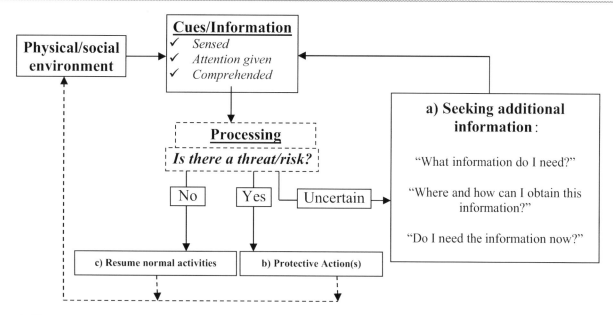

Fig. 4.2 The protective action decision model, redrawn and adapted to building fires

4.2.1 Cues

The first stage of the PADM, shown in Fig. 4.2, is the presence of cues or information from the physical or social environment during the building fire event. This section will aid the engineer in understanding the types of cues that they need to account for when performing egress and the ways in which these cues may impact the actions taken in the pre-evacuation and movement time periods of the evacuation timeline.

Cues can be provided by all three elements of a building fire event, as is shown in the following list:

- Cues from the fire environment (or fire cues)
- Cues from the building
- Cues from people in the building

Fire cues, such as heat, smoke, or crackling sounds, are generally present in the area of fire origin, but may also be present in other areas of a building due to the spread of smoke and/or flame. Fire cues may be a first cue to some occupant groups and may trigger the initiation of other cues such as people alerting others.

Cues can also originate from the building itself. Automatic cues may result from building signaling or public-address systems. In the case of fire alarm systems, the presence of automatic detection and notification equipment needs to be established. The type, capabilities, extent of coverage, and reliability of any notification system needs to be considered in the context of the occupant group (s). Building service disruptions, such as power failure, can also create cues for occupant groups in areas near or remote from a fire location. For occupants remote from the fire, such cues may be disregarded but may also begin the protective action decision-making process. Such cues may be important in this process, but may not readily initiate evacuation.

Additionally, there can be cues provided by people in the building, such as verbal staff instructions, observable occupant actions, and fire department arrival. The effectiveness of visual, verbal, or tactile cues from building occupants to prompt other occupants varies with occupant characteristics. Some types of occupants may be more readily influenced by alerting actions while others may not.

> **Cues can vary by type and intensity in a building fire**
> In a study of the 2001 World Trade Center Disaster, building occupants in both WTC 1 and WTC 2 encountered cues that varied in type and intensity throughout the building evacuation timeline [51]. The table, below, taken directly from the report on "Occupant Behavior, Egress, and Emergency Communications" [51] shows a summary of the ways in which survivors of WTC 1 became aware that something was wrong. The plane strike was immediately obvious to occupants in WTC 1—both by feeling something (63%); such as the building moving, shaking or swaying; and hearing something (30%); such as a boom, crash or explosion sound. Other WTC 1 survivors (7%) became aware that something was wrong by seeing the plane, smelling jet fuel, being knocked off of their chair, and/or being warned about the event by someone else.

How survivors in WTC 1 became aware something was wrong on September 11, 2001

	Percent (n = 44, %)
Felt something (building moving, impact, shaking, swaying)	63
Heard something (boom, crash, explosion, blast, roar, rumbling)	30
Other, including saw a plane, smelled jet fuel, fell down/fell off chair, warned by someone	7

The engineer should be aware that not all cues provided to the building occupant are relevant and/or relate to the fire event. In some types of events, cues are provided that have nothing to do with the fire event but can distract or interrupt the occupant decision-making process.

4.2.2 Sensing the Cue(s)

At some point during the fire event, people will likely begin to sense (or receive via any of the five senses) cues from the fire environment, the building, and others in the building. However, it is important to understand that sensing the cues may not happen immediately after ignition, and in some cases, may not happen at all, which would significantly delay the decision-making process.

There are five ways in which individuals can sense cues from their environment. Those five ways are through sound (i.e., hearing), sight, touch, smell, or taste. Depending upon the type of cue, individuals can receive cues from a fire event in various ways.

4.2.2.1 Fire Environment

Fire cues, i.e., those related to the fire and combustion, can be received via multiple senses. The fire cues include olfactory cues such as the smells of smoke, burning, plastic, acrid, etc. There are also visual cues, such as the visible smoke that can be any shade from white to black and of different density. Other visual cues can also be sparks or flames. The fire can also emit auditory cues such as crackling sounds. The heat from the fire is also another potential cue that could be felt through the skin.

Few people are actually awakened from sleep by an odor stimulus

In a study of waking effectiveness of smoke odor, only 20% of the subjects tested during their sleep (at stage 2, which is marked by muscle tension and gradual decline in vital signs) were awakened by a smoke odor [52].

In a study of 17 young, adult volunteers with ages ranging from 18 to 26 (mean = 21.4) with self-described "normal" smell and sleep patterns, 29% of males and 80% of females woke when exposed to a smoke odor. The smoke odor concentration at the pillow was approximately 1 PPM at 1 min and 6 PPM at 10 min. The subjects that awoke in response to the odor did so between 45 and 205 s after emission of the odor began, with a mean time of 101 s (SD = 56 s) [19].

Auditory cues may actually be more effective at waking sleeping adults

Waking patterns to light and auditory cues, similar to those that might be emitted by a fire, were tested in a sleep study involving 33 adult volunteers with normal hearing and sleep patterns, aged 25–55 (mean = 43 years). Ninety one percent woke to a crackling sound, and 83% awoke to a shuffling sound. The sound was in the range of 42–48 dBA at the pillow, and 83% of the people who woke up in response to the cue did so within 30 s.

The light source used in the experiment was in the range of 1–5 lux. Forty nine percent of the subjects awoke in response to the light cue, and 50% of those did so within the first 30 s [19].

4.2 The Protective Action Decision–Making Process

> **The effects of irritant smoke can inhibit our ability to sense other cues from our environment**
> Irritant smoke in the eyes or nose immediately causes pain (sensory irritation), reflex closure of the eyes, and breathing difficulties. In a test of human subjects, Jin et al. found that the presence of irritants in smoke significantly affected visibility and movement; subjects had to continually close their eyes due to the irritation, and tear production in the eyes caused obscuration of exit signage. In an experiment, Jin et al. found that subjects exposed to irritant smoke also experienced emotional instability and thought process decrements. When polled on the "uncomfortable factor of fire smoke", participants consistently ranked smoke irritation in the top three causes of discomfort [53].

4.2.2.2 Building Characteristics

Cues from the building vary in the way that building occupants can receive them. First, a building's signaling system could include audible alarm signals such as bells, chimes, horns, and automatic voice messages, as well as visual alarm signals such as strobes, flashing signs, messages on information screens, etc. Receiving the building signalling cues can depend on the perceptual characteristics of the cues. The audibility and visibility of various signals will determine if occupants can sense these cues, as shown in the case studies provided below.

> **Characteristics of the fire alarm, including sound pressure level and location within the room, and characteristics of the person, including age, can influence the time to awake from sleep**
> In an experiment conducted with 36 subjects aged between 6 and 59 with normal hearing and sleeping patterns, 100% of adults (aged 30–59) awoke to a 3-min, 60 dBA (measured at the pillow) smoke alarm, while 85% of children (aged 6–17) slept through the alarm [54].
>
> An experiment with subjects aged 19–29 examined sleep-waking performance in response to fire alarm signals and environmental noise [55]. Alarm sound pressure levels of 70 to 85 dB levels (at head of bed) awakened 20 of 20 subjects. The average waking time in response to a sound pressure level of 70 dB and 85 dB was 9.5 s and 7.4 s, respectively. Alarm sound pressure levels of 55 dB awakened 80% of the subjects, and the average waking time was 13.6 s.
>
> *Sound power and sound pressure levels are expressed in decibels (dB). See the SFPE Handbook of Fire Protection of Fire Protection Engineering [2] and NFPA 72® for more information [56].

> **People do awaken from strobe lights; however, the effectiveness can depend on the sleep stage of the individual. Additionally, intermittent bed shakers can wake people from various stages of sleep**
> A study investigated waking in response to light [57]. Thirty normally hearing college students and 48 deaf subjects were used, and the light sources were strobes, industrial strobes and 100-watt incandescent lights. White flashes were perceived as the brightest, and the study concluded that "clearly, signal intensity is a major factor in arousing sleeping people."
>
> | | Household strobe | Industrial strobe | Incandescent |
> |---|---|---|---|
> | Deaf subjects awakened (%) | 86 | 92 | 92 |
> | Hearing subjects awakened (%) | 82 | 78 | 59 |
> | Deaf subject mean waking time (s) | 16 | 15.9 | 31.2 |
> | Hearing subject mean waking time (s) | 23.9 | 13.4 | 53.5 |

Ashley et al. evaluated the waking effectiveness of audible, visual, and tactile devices on subjects with various hearing levels. [58] The subjects were tested during active Stage 2, Delta, and REM sleep. Researchers found that strobes were the least effective in awakening hearing impaired and normal hearing individuals. The waking effectiveness of the strobe was dependent on the sleep stage of the individuals. For normal hearing subjects, the effectiveness of the strobe varied from 25 to 40%. For hard of hearing individuals, the effectiveness of the strobe varied from 17 to 71%, and for the profoundly deaf, the effectiveness of the strobe varied from 42 to 75%. In all cases, the intermittent bed shaker was 100% effective in awakening the full range of test subjects. As would be expected, the audible smoke detector did not awaken any of the profoundly deaf subjects and was only effective in awakening 55–62% of the hard of hearing population.

Additionally, cues from building services disruption can be received audibly, visually, or via tactile or olfactory means. At the same time, building characteristics can have an impact on the delivery of cues to occupants. Factors such as ambient noise, odors and light, degree of compartmentation, ceiling heights, ventilation conditions, etc., can enhance or deter cue delivery.

4.2.2.3 People

Cues from other people include verbal alerts as well as visual observations of the behavior of other people, and even in some cases, tactile alerts (i.e., someone physically alerting someone to danger). Such cues obviously will only be available in spaces where other occupants are located or through which other occupants will pass.

Another aspect of the "people" in a building fire is the factors of the "receiver" (i.e., the individual receiving any type of cue), that will ultimately influence whether they sense the information. This should not be confused with the previous paragraph that talks about the actions of others (i.e., the senders of the cues). The factors that can inhibit the sensing of cues by the "receiver" include the presence of sensory impairments, such as being hearing or visually impaired. There is also variation in sensitivity to smells and heat as well as differences in ability to see the smoke.

Individuals may also be so involved or committed to certain pre-emergency activities that they may miss (or simply do not sense) cues indicating an emergency is taking place. This is often referred to as "commitment." For example, a person working intently at the computer might not see smoke filling up a room while a person entering that room could readily perceive the unusual cues.

4.2.3   Paying Attention to the Cue(s)

The next stage in the decision-making process is whether or not the individual will pay attention to cues provided by their environment. Simply, because someone hears, sees, smells, tastes, or feels any type of cue from a fire event, does not necessarily mean that they will pay attention. And, in cases where cues are ignored, the decision-making process is delayed, which in turn, lengthens the pre-evacuation period of the building evacuation timeline. In cases where the fire environment is evolving for the worse, a delay can decrease the chances for survival.

Frequently, this stage of the decision-making process may overlap or seem to overlap with the sensing stage (1b) of this process. A line is drawn between these two stages here to identify additional factors that may hinder this process, but it may be more useful for engineers to see the two stages as one in the same.

Regardless of the cue type (e.g., environmental, building-related, or people), there are certain factors that can influence whether a cue is given attention. For example, cues that are located in someone's field of perception (e.g., line of sight or within earshot), provided with sufficient contrast from the background or environment (e.g., visual color contrast or different speaking voice), and/or presented in an obvious manner (e.g., visually larger, audibly louder, or with more sensitivity to the touch) are more likely to grab a person's attention in a fire situation.

Examples are provided here by cue type. One example of a building-type cue can be an audible message provided via the building's public-address system. The presence of a verbal warning message that is disseminated so that all affected building occupants can hear the message, provided by a different voice than usually provided messages to the building occupants, and given with a greater sense of urgency or at a higher volume than normal, can grab attention. Another example is the actions of a fire warden in the building. A fire warden that physically travels to each building occupant's field of perception, speaks or acts in a manner that is different from normal, and then speaks in an urgent tone or performs actions urgently or with purpose, can grab attention.

Additionally, there are certain states in which "receivers" may find themselves at the time of an emergency that can hinder their ability to pay attention to cues from the fire event, including drug/alcohol impairment, cognitive impairment, sleep, and/or sleep deprivation [59, 60], or in some cases, facilitate their ability to pay attention to cues, including hypervigilance. A case study is provided below as an example. Also, disaster-induced factors can inhibit attention paid by "receivers" to emergency information. Stress and anxiety brought on by the emergency can narrow an individual's focus on certain details, and once the focus is narrowed, unexpected features are simply not noticed (known as inattentional deafness or blindness) [61, 62]. Finally, if individuals are part of a larger group, they can be less likely to pay attention to their external environment.

> **Alcohol consumption increases the fire fatality risk for people in all age groups**
> This is an important issue especially for young adults, who tend to be less experienced drinkers and deeper sleepers. Ball and Bruck [60] investigated the effect of alcohol on the arousal threshold for three alarm signals.
>
> The participants in this study were 12 students (seven males, five females) in the ages 18–25 years old from Victoria University. The minimum sound volume was sought at which students would awaken to any of three alarm signals when they were sober or had a blood alcohol concentration (BAC) of 0.05 or 0.08 as measured with a breathalyzer. The first alarm signal was a female voice warning of danger from fire and instructing the subject to wake up and investigate. The second alarm was a high frequency signal from a standard Australian home alarm. The third alarm was a low frequency Temporal-Three (T-3) pattern from ISO 8201. Subjects were tested on a different night for each level of intoxication with three to seven days between trials. The alarm signal was played during the deepest stage of sleep (stage four) in each participant's bedroom, with a background noise level of about 50 dBA. To familiarize participants with the alarms, all signals were played for them before they went to bed. Each alarm was initially sounded at 35 dBA for 30 s and then increased by 5 dBA until 95 dBA was reached, at which point the alarm was allowed to sound for 3.5 min. Results are shown in the Table below. The female voice and T-3 signals were equally effective, and significantly better than the standard alarm in waking the subjects. Subjects were more likely to sleep through the alarm as alcohol level increased, with 36% of trials at 0.05 BAC resulting in no response below 95 dBA. The sex of the student was not a significant factor.
>
> Sound levels required to awaken people who are sober or under the influence of alcohol.
>
> | Alarm Signal | Mean Threshold Waking Level (dBA) | | |
> |---|---|---|---|
> | | Sober | 0.05 BAC | 0.08 BAC |
> | Female voice | 59.6 | 79.3 | 80.5 |
> | Standard alarm | 72.5 | 85.0 | 87.5 |
> | T-3 pattern | 59.2 | 78.3 | 82.9 |

4.2.4 Comprehending the Cue(s)

The third step in the decision-making process is comprehension, or understanding the information being conveyed by the cue. This stage often involves the cognitive activity of comparing the cue and/or information to knowledge, experience, or training from the past. A person may correctly interpret a cue for what it is, mistake the cue for something else, or not understand the meaning of the cue at all. In cases where the cue is not understood or mistaken for another, the engineer should note that this may delay the decision-making process, and in turn, lengthen the pre-evacuation time period of the building evacuation timeline.

4.2.4.1 Fire Environment

Comprehending fire cues will depend on the "receiver's" knowledge and past experience with similar fire cues. Having been through a fire event, or simply having experienced very controlled fires in fireplaces or bonfires, forms part of the knowledge people have about fires. A person does not need to have been through an actual fire to understand fire cues; however, it would aid in their comprehension of cues from the fire environment.

4.2.4.2 Building Characteristics

Prior exposure to a specific alarm system or to alarm systems in general leads to learning. Based on prior training and experiences, a person will be more likely to understand the meaning being conveyed by an alarm signal [44, 63]. Occupants or staff who have been trained for fire response through fire drills, should more readily recognize fire alarm signals than other building occupants.

In a public building when a fire alarm sounds, occupants may not necessarily comprehend the signal as a fire alarm because there are a variety of warning signals in the built environment and it is very difficult for the public to discriminate these different signals [64].

> **Occupants will not recognize an alarm signal, even the temporal–3 pattern, without proper training and education**
>
> Proulx et al. [65] performed research to assess the public's recollection, identification, and perceived urgency of the fire alarm signal, known as the temporal-3 pattern. This alarm pattern was intended to become the standardized alarm signal around the world; however, no formal public education had taken place to inform building users of the meaning of the alarm, and the subsequent response expected of them.
>
> This study was conducted in Canada with 307 visitors of public buildings, who had not received any training regarding the temporal-3 pattern evacuation signal. Results showed that only 6% recognized or associated the temporal-3 pattern to a fire or evacuation signal. Though the temporal-3 pattern signal can be used with any sound for the evacuation signal, including all of the sounds used in the study, only a single electronic tone having a fundamental frequency of 505 Hz was used for purposes of evaluating recognition of the temporal-3 pattern. Other sounds tested included a car horn (98% recognition as a car horn), vehicle back-up alarm (71% recognition as a back-up alarm), bell (50% recognition as a fire alarm), slow whoop (23% recognition as a fire alarm) and a buzzer (2% recognition as an industrial buzzer). On an arbitrary scale where people judged the perceived urgency conveyed by the signals, the temporal-3 pattern signal ranked the lowest. Results show that public education is essential to increase the perceived urgency and recognition of the temporal-3 pattern for occupants to readily identify the evacuation signal [65].

Building services disruption to a fire event, such as power outages or breaking glass, may not be comprehended initially as a fire cue. Instead, the building occupant or occupants will likely require the receipt of and attention paid to additional fire cues to move on to subsequent stages in the decision-making process.

4.2.4.3 People

The ability to comprehend cues provided by others, especially in the form of warning information, will depend in part on the language used in the warning. The use or overuse of complicated terminology, jargon, acronyms/abbreviations, or an overload of information, for example, will increase the difficulty for occupants to comprehend the information.

Additionally, factors associated with the "receivers", themselves, can inhibit comprehension. These factors include being untrained or unprimed, age (children), being a non-native speaker (especially those who do not speak the native language at all), having a cognitive impairment, and being from different cultural backgrounds. Even more complicated is the influence of stress (induced by the emergency) on the general public's ability to understand emergency information. Verbal comprehension of emergency messages, for example, has been found to drop an average of four grade levels at high levels of stimulation, stress, and distraction (all factors that are possible during a building emergency) [66].

4.2.5 Processing the Cue

At this stage in the decision-making process, the building occupant will attempt to assess the presence of a threat and the risk to him/herself or to others. The assessment of a cue (or cues) involves two different interpretations. The first is known as threat assessment, for example, "it is truly a fire?" The second is risk assessment, for example, "it is threatening to me?"

Unfortunately for building occupants, cues from fires can vary significantly, thus the information available to occupants during a fire event is often characterized by its ambiguity. Smelling smoke, recognizing the sound of an alarm, or observing behavior of others/staff may not indicate to occupants that "there is an actual fire that requires evacuation." Generally, the initial information available to define the situation is perceived as ambiguous because it is often contradictory, unusual, and unexpected.

Because of this, threat and risk assessment may or may not be accurate, and it may also change over time as new cues are processed. This stage in the process will determine the protective action response applied by the person. Note: If the occupant does not decide to take protective action, they are essentially extending the time he/she spends in the decision-making process, which in turn, lengthens the pre-evacuation time period of the building evacuation timeline.

4.2.5.1 Fire Environment

Certain characteristics of fire cues are more likely than others to prompt individuals to determine that the threat and risk are credible during a building fire event. A larger number or greater intensity of the fire cues (e.g., the thicker or the darker the smoke, the more intense the fire cue), can influence an individual's threat and risk assessment.

4.2 The Protective Action Decision–Making Process

More intense cues lead to shorter delay in a building fire

A study of the evacuation of the World Trade Center in New York City following the 1993 bombing showed that the mean delay times (or pre-evacuation times, as labeled in this chapter) were significantly shorter in the North Tower, where most surveyed respondents reported feeling or hearing the explosion, than in the South Tower, where the cues available to the occupants were more ambiguous [67]. Further, the South Tower occupants who first became aware of the incident because of the explosion, as opposed to the loss of electrical power to the building, were significantly more likely to believe the situation was extremely serious.

However, the threat and risk assessment stages may not go hand-in-hand. Just because an individual considers the situation to be a credible threat, does not necessarily mean that he/she will personalize the risk, and in turn, respond to the event. For example, an individual may see smoke in a building and assess the situation as a credible threat (i.e., a real fire event is taking place); however, he/she may begin to engage in a firefighting response, rather than move to safety, since he/she does not necessarily feel at risk from the fire event.

4.2.5.2 Building Characteristics

Similarly, just the sounding or visualization of building signalling systems does not guarantee that individuals will believe that a real fire event is taking place and that they are at risk of danger.

Fire alarm bells, alone, will not convince an occupant that they are in danger

Over 36% of survivors of a high-rise residential fire, which killed six people, mentioned that their initial interpretation of the situation after hearing the alarm bell and perceiving smoke was that the situation was probably "not at all serious" [68].

The likelihood that a warning system will be interpreted as intended (i.e., as a real event) depends on its credibility, and its credibility can be enhanced by disseminating an alert and warning message. Chapter 12, Enhancing Human Response to Emergency Notification and Messaging, can be consulted for further information on the dissemination of effective alerts and warning messages.

Providing detailed information about the fire event will be more convincing to people that there is a fire occurring and that they are in danger

Ramachandran conducted two experiments to determine which types of alarm notification are most likely to be interpreted as a fire. The results indicated that providing more detailed information, such as by voice, graphic display, or text, resulted in as much as a six-fold increase over a bell of interpreting notification as indicative of a fire [69].

The credibility of a building signal can also be reduced by the continual presence of false alarms. In other words, the alarm/warning system may lose credibility if it has initiated multiple times in instances when no emergency has materialized or occurred. Breznitz, in his book *Cry Wolf Syndrome* [70], states that the detection component of a warning system is electrical or mechanical; the management of information gained, as related to occupant notification, must consider the psychological aspects. Breznitz further states that in order to avoid potentially irreversible negative consequences of false alarms, one can design warning systems to better manage the signal.

False alarms lead to lower confidence in fire alarm systems

A false alarm study was conducted in a college dormitory. In addition to recording numbers of false alarms, the overall effect these had on "people response" was observed. "Such smoke detection systems, together with prevailing resident attitudes, may offer no life safety improvement over simple manual pull stations and local room, residential smoke detectors and may actually render the buildings environment less safe because the more numerous alarms are more likely to be ignored with systems smoke detectors present. A false alarm rate of approximately one per week caused students to believe that there was no "credibility status" for the alarm system [71].

Lower confidence in the fire alarm system leads to longer response delays
Two hundred thirty-eight unannounced fire drills were conducted in U.S. Veterans Affairs facilities in the United States [63]. An experimental procedure for measuring health care staff "response delay" and a 48-question participant survey was used to gather information for a multiple regression analysis. Correlations between the independent variable, "Time to action", and the 48 factors revealed two findings with moderate correlation coefficients:

Larger fire alarm zone systems correlated with longer staff response delays and lower confidence in the fire alarm system correlated with longer staff response delays.

The following factors also correlated, in order of importance, but explained a very small percentage of variance in response delay:

- Coded signal systems correlated with longer response delays
- More fire protection systems and equipment within the zone correlated with longer response delays
- More alarms over the past 6-months (all alarms) correlated with longer response delays
- More false alarms over past 6 months correlated with longer response delays
- If staff person's location involved direct patient care, there were longer response delays.

Response delay times were measured for the fastest, or first responding staff person within the zone alarmed. The staff in the study were highly trained with a high sense of responsibility to react quickly and properly.

Similar to other types of building cues discussed above, building services disruption that occur often, especially in non-emergency situations, can decrease the likelihood that threat and risk assessments will occur. Instead, they may hinder the process, and even influence individuals to believe that no threat (or fire event) is occurring and that they can go back to their previous activities.

What may actually prompt individuals to believe that an actual fire event is taking place is observing or witnessing the arrival of fire trucks and/or emergency (or fire department) personnel to the building. An example of the effect of witnessing fire trucks and/or fire department personnel is provided as a case study, below.

Building occupants can become convinced that an actual emergency is taking place after witnessing emergency response vehicles at the scene
In the 1993 bombing incident at the World Trade Center in New York, several respondents reported that their initial awareness of the situation was hearing the sirens of the responding fire apparatus [67].

4.2.5.3 People

Interpreting cues from others will depend in part on the credibility of the people generating the cue. A person who repeatedly raises a false alarm will likely not be credible. For example, if a neighbouring worker is prone to become agitated over trivial events, the receiver may not heed any cues and in turn, not recognize a warning for what it is. Unusual activities, such as general movement toward exits, or increased levels of conversation mentioning a perceived threat might also influence others to believe that something is going on and/or that they are in danger. On the other hand, the failure of others to take action can have an inhibiting effect on building occupants' threat and risk assessment.

People have a reluctance to interpret cues as threats if others around them do not
A study done at Columbia University in the 1960s clearly illustrates this phenomenon [13]. Subjects were seated in a waiting room, ostensibly for an interview, and while they filled out a questionnaire, smoke was introduced into the room through a wall vent. Subjects who were in the room alone were significantly more likely to leave the room to report the smoke than those in the room with two others, either other naïve test subjects or 'actors' participating with the research team. Interestingly, in interviews after the experiment, subjects who had reported the smoke were likely to mention that they had considered the possibility that there was a fire, while those who had not reported the smoke all said that they had rejected the idea that there was a fire.

4.2.6 Decision–making and Taking Protective Action

In the decision-making stage of the process, occupants have to decide if an evacuation is required or necessary, and if so, what to do next. It is important to understand that the decision-making process takes time to complete. Under environmental-, building-, and people-related conditions, described below, these decisions may not always result in the safest, most effective actions, which can lengthen the pre-evacuation time period of the building evacuation timeline. And, even when decisions are made that protective action is necessary, other actions may be taken (e.g., to protect property or others) that can lengthen the time spent in the pre-evacuation time period.

Janis and Mann [72] have explained that decision-making during an emergency is different from day-to-day decision-making in two main ways; "One is that there is much more at stake in emergency decisions - often the personal survival of the decision maker him/herself and of the people he (or she) values the most. A second important difference is in the amount of time available to make a choice before crucial options are lost."

Under time pressure and/or distracting conditions, according to Wright [73], decision makers are inclined to use less information to make decisions. In other words, not all possible information and alternatives are considered [74]. The use of only a few units of information is supported by heuristic models that explain the person's need to reduce the space of the problem to an easily manageable dimension [75]. The decision is taken according to Simon's concepts of "bounded rationality" and "satisficing"; when a good-enough solution for the problem is found [76].

In this stage, the individual (or group of individuals) is essentially asking the following question: "Do I need to take protective action?" If they assess that there is a credible threat, and they perceive a certain (credible) level of risk related to his/her own safety or the safety of others, then the individual or group of individuals will decide that protection is necessary. In a building fire, this often involves the decision to evacuate to a place of safety, but also could include a decision to remain in place, if that is the safest option. After the decision is made to seek protection, the final step in the decision-making process involves the implementation of a protective action plan or strategy.

The protective action plan or strategy often involves the performance of a variety of actions to protect property (or prepare to take protection), protect others in the building, and/or protect oneself. Examples of each of these action types are presented in Table 4.1.

As shown in Table 4.1, to protect property or prepare for taking action, building occupants often gather belongings, get dressed or put on outerwear, secure items from fire, and/or fight the fire, among other actions. Additionally, there are actions that can be performed to protect others, such as the following: physically assisting others to safety, instructing others to take protection (e.g., evacuate), or searching for others. Occupants have also been observed re-entering buildings previously evacuated to protect property or others. Finally, occupants perform actions to protect themselves, including evacuating, moving to a place of safety somewhere else in the building, or remaining in place until further instructions are given. Another decision of importance related to protecting oneself involves deciding which route to take to achieve safety.

Just as in other stages, there are factors (i.e., environmental-, building-, and people-related), that can inhibit certain decisions related to the protective action plan. This document has already mentioned the influence of a few environmental factors, such as time pressure and stress, on decision-making, but has not yet covered the influence of the three cue types on decision-making during fire evacuations.

4.2.6.1 Fire Environment

Narcotic gases that are produced from fire events can influence decision-making during fire events. Studies have focused on the effects of sublethal doses of asphyxiant gases, including carbon monoxide (CO) and hydrogen cyanide (HCN) on people and animals. However, it should be noted that we cannot predict the effect of these gases on decision-making and/or response with extreme accuracy.

Table 4.1 Examples of protective actions performed during building fires

Protect Property (prepare for action)	Protect Others	Protect Oneself
Gather belongings	Physically assist	Evacuate or leave the building
Get dressed/outwear	Instruct others to take action/warn others	Move to a place of safety within the building (relocate)
Secure items from fire	Search for others	Remain in place
Fight the fire	Contact / inform authorities	
	Inform family/friends	

At higher levels (above 20%), COHb can affect occupant decision-making and taking action

The overwhelming majority of the data available from human subject testing on visual impairment, auditory impairment, and reaction time suggests that there is little to no effect when carboxyhemoglobin concentrations are at or below 20% [77]. Researchers found that schedule-controlled behavior in laboratory animals (e.g. rate and pattern of responding under schedules of reinforcement such a lever pressing upon food presentation) was not affected until carboxyhemoglobin (COHb) levels exceed 20% which is consistent with the human subject testing; the main effects were on the rate of processing and/or responding to cues.

Note: Smokers and individuals who live in heavily polluted areas will have a higher baseline COHb than those individuals who are not exposed to habitual and environmental sources of carbon monoxide. Those individuals with higher baseline COHb levels (e.g. 3-8% as opposed to 0-1%) will require less time to reach a 20% COHb level than an individual who has not been pre-exposed to carbon monoxide [77].

Above 20%, exposed subjects begin to experience headaches, difficulty breathing during exercise, and other minor physiological changes. As blood concentrations approach incapacitating ranges (e.g. 30–40%), however, exposed subjects can experience confusion, disorientation, and loss of judgment which are all critical factors in evacuation and escape.

In another case study, an unconscious fire victim was found by the fire department just inside the front door of her apartment [78]. She was promptly transported to the hospital and treated for carbon monoxide exposure. Upon recovery, the victim recounted that she was attempting to open the front door to escape but could not complete the task of turning the door knob. Eventually, she became unconscious at the doorway from carbon monoxide exposure. The victim recounted that she had early notification of the fire but attempted to extinguish it and locate one of her missing children prior to evacuating the apartment. The victim's COHb level was back calculated to account for oxygen treatment and wash-out and was found to be approximately 25%.

Exposure to HCN can also negatively influence actions taken during a building fire

Purser et al. [79] investigated the sublethal effects of hydrogen cyanide exposure in monkeys. Purser et al. found that the effects of exposure to low level of hydrogen cyanide gas when compared to exposure to burning polyacrylonitrile (which has a high nitrogen content) were the same. The monkeys experienced loss of consciousness after 1–5 min of exposure followed by cardiac disruption. The results of testing strongly suggested that hydrogen cyanide could produce incapacitation at low levels during which time the victim would continue to experience exposure to lethal levels of carbon monoxide.

However, there are success stories from fires, where people were able to walk through smoke to evacuate the building. Therefore, the engineer should not automatically assume that building occupants will avoid movement through smoke. Instead, studies conducted with fire survivors in homes demonstrate that a large number of occupants moved through smoke to escape [30].

People will move through smoke during evacuation

In the UK interviews with 2193 people involved in fire incidents show that 60% moved through smoke; a similar study in the USA with 584 people involved in fire incidents shows that 62.7% said they moved through smoke – even at occupant moved some distance through smoke when the reported visibility distance was less than 0.6 m (or 2 ft) [30].

Although occupants were prepared to move through smoke, many eventually decided to turn back because of the smoke, heat or a combination of both. For the UK population, 26% made the decision to turn back at some point during their evacuation and the same decision was made by 18.3% of the USA sample.

Studies have also been conducted to understand how occupants of non-residential structures interact with the fire environment. Movement of occupants through smoke was witnessed in behavioral studies of the 1980 MGM Grand fire [80], the 1993 WTC bombing [67], the 2001 WTC disaster, the 2003 Cook County Administration Building fire [81], the 2003 Station nightclub fire [82] and evacuation through tunnel fires [83], among others.

4.2.6.2 Building Characteristics

Similar to how cues can affect decisions made by building occupants, so can characteristics of the building itself. Building characteristics have been shown to influence route choice, in that exits that provide more affordances are more likely to be used. Nilsson brings the Theory of Affordances to fire research in his study of the characteristics of emergency exit doors that make them more attractive to building occupants [84]. According to the Theory, emergency exits are more likely to be used if they are one or more of the following: (1) easier to see, (2) easier to understand (i.e., how to use), (3) easier to use (e.g., unlocked or requiring a smaller force to open), and helpful in allowing occupants to achieve their goals (e.g., lead directly outside of the building for evacuation purposes).

> **Several experiments have been conducted at Lund University in Sweden to study effective design of emergency exits**
>
> In one series of experiments in a cinema theater, green flashing lights were used to mark emergency exits to see if that would influence exit choice [18]. In two of four unannounced evacuations in the theater, green flashing lights were installed at one of two exits. Study results showed that in the experiments where the flashing lights were used, all subjects used the emergency exit, while in the experiments where the flashing lights were not installed, no one used the emergency exit. In another experiment at an IKEA store, it was observed that occupants were unlikely to use any emergency exit, but when they did use an emergency exit it was an exit that they walked toward during their evacuation; they generally walked right past another emergency exit that was parallel to their walking direction [85].

4.2.6.3 People

The actions of others have been shown to influence individuals' (i.e., the receiver's) decision-making in response to a fire event – both the actions that the receiver performs, and the routes that the receiver chooses to take to reach safety.

The actions that the receiver performs can be affected by what others are doing in the building. If others in the buildings are not taking action, this may influence the receiver to refrain from taking action as well (even though he/she may wish to act). Fear of social embarrassment has been found to influence protective action in various types of research settings (e.g., Latane and Darley [13]) [86]. On the other hand, if individuals are taking action, or even following certain exit routes from the building, this may influence receivers to follow suit.

Social affiliation between the receiver and other occupants in the building is another characteristic that predicts when receivers are likely to attempt to protect others (i.e., notify and gather with people with whom they have emotional or social ties, such as family members or social or business groups). This activity of notifying or gathering members may take time, especially if members are not together at the initial awareness of the fire incident. These activities may also involve movement toward the fire area and through smoke to gather missing members.

> **Social ties with others have been shown to lengthen time spent in the pre–evacuation period**
>
> The social affiliation behavior was well documented by Jonathan Sime who studied, among other cases, the behavior of the occupants of the Marquee Show Bar, part of the Summerland holiday leisure complex in England, where a fire killed 50 people in 1973. From interviews, it was determined that occupants in the bar who had left their children in another area of the complex started to move rapidly after the initial cues of a fire. Moving at counter flow to the people evacuating, they attempted to find their children. Groups who were intact in the bar took a longer time to react and to start leaving the premises, putting the whole group in jeopardy. This was because where the group was not intact, members of the group decided to get the missing individual(s) [16].

Additionally, the characteristics of the receiver, namely, their role in the building, their familiarity with the building, their commitment to specific activities, and their mental and physical states and capacity, can all influence the actions that they perform and the routes that they choose, personally.

The receiver's role in the building and/or this person's feeling of responsibility toward others and/or the building itself will likely influence his/her decision to take action, as well as decisions on what actions to take. For example, in a single-family home, occupants recognize that they are responsible to initiate adaptive action on their own. On the other hand, in a public building, such as a museum or a shopping center, visitors may not assume the responsibility to initiate adaptive behavior if the fire alarm is activated, whereas staff or even building management will likely initiate adaptive action on their own as well as assist in others' protective action plans. Consequently, an occupant's role is an important factor to consider when examining their decision-making process and protective action plan during a fire incident. Occupants who are visitors might simply make the decision to wait while staff may decide to investigate the problem and/or provide occupants with definitive information.

> **Staff of businesses can be considered a credible source, and thus influence others to evacuate, during a fire event**
> The FireSERT research group at the University of Ulster conducted unannounced evacuations in Marks and Spencer's department stores with neither staff nor customers aware of the exercise. One, for example, was recorded through 46 video cameras and followed with a questionnaire of approximately 300 customers. When the fire alarm bells sounded, customers looked around at the behavior of others. It is only when staff started to close their cash registers and instructed the customers to evacuate that the occupants complied and moved toward exits. Among the evacuees interviewed, 52% said they were prompt to evacuate when requested by staff. The mean pre-evacuation time for this evacuation was 24 s. This evacuation demonstrates the importance of staff training and the influence staff can have on decision making and evacuation during an emergency in a public building [87].
> In the Beverly Hills Supper Club Fire, the wait staff remained in their roles, assisting in the evacuation of patrons at "their table." [88]

Another factor that can influence actions performed, in this case, route choice, in an occupant's familiarity with the building. As mentioned earlier, it is expected that a person will not refer to all possible information and alternatives when making a decision. Only a few options, which the subject considers as being more likely to solve the problem, will be retained. A rapid and easy strategy to make decisions and solve a problem is to apply a well-run decision plan [89]. Usually, for the person, this translates into evacuation by a familiar route [16]. In public buildings, this usually means leaving by the main entrance [16]. Only when the familiar route is blocked or judged dangerous, will the person try to develop an alternative plan or strategy.

> **People are likely to use the main exit (i.e., the exit with which they are most familiar, for evacuation)**
> A number of evacuation drills have been conducted with regular, non-informed customers in different IKEA stores in Sweden. It was observed that shoppers attempted to exit the stores either by the entrance or by the regular exit moving along the zigzag path that took them through different departments before reaching the exit. Customers passed by several emergency exit doors without using them as long as they were not open [85].

Commitment of the receiver is one of the most important determinants of decision-making. If a receiver is committed to a specific activity, this might aid an individual in his/her decision to refrain from taking protective action. In other words, depending upon the perceived importance of the initial activity, a decision might be made to ignore the situation until sufficient information requiring a decision is imposed on them.

> **Individuals were committed to activities in the 1987 King's Cross fire that likely put them in harm's way**
> In November 1987, a fire erupted on an escalator killing 31 people at the King's Cross underground station in London, England. The fire and smoke spread in the escalator area, the ticket hall and toward the different corridors and entrances to the station. It was observed, during that fire, that passengers continued their routine activity of traveling home; they entered the smoke-filled station, went down escalators, sometimes next to the visible flames, in an attempt to catch a train to go home. These people were committed to use the underground to reach a destination and were unlikely to shift their attention on ambiguous circumstances for which they felt they had no responsibility [90].

Mental alertness and limitations are other characteristics, which, with age, may limit the receiver's capacity to react in a given situation. For example, if a fire starts in the middle of the night, occupants who are not alert because they are asleep will require a longer time to respond, or may not respond at all, if not alerted by others. Another dimension is the possibility that occupants may have some limitations that will extend their decision and response time. These limitations could be perceptual, physical or cognitive or might be due to the consumption of medication, drugs, or alcohol. Finally, very old and very young occupants may have limited physical or cognitive capacity to perform certain protective actions. However, it should be noted that we cannot predict the effect of these limitations on decision-making and/or response with extreme accuracy.

4.2 The Protective Action Decision–Making Process

Alcohol is detrimental to the successful escape of victims from fires

Studies have shown that alcohol intoxication is the precursor for many fire related deaths [91–96]. Barillo et al. [96] compared data on victim locations and blood alcohol concentration (BAC) and found that those victims with an average BAC of 0.268 g/dL were found in their bed and showed no signs of escape. Those victims with an average BAC of 0.116 g/dL were found between the bed and an exit, and victims with an average BAC of 0.088 g/dL were found at the exit. Levine et al. reported a variety of effects on performance based on blood alcohol level. In the range of 0.09 and 0.25 g/dL, loss of critical judgment, impaired perception, memory, comprehension, and decreased sensory response is found, amongst other things. As the alcohol range increases from 0.18 to 0.30 g/dL, disorientation, mental confusion, dizziness, and disturbances in vision and motion are among the intoxicating effects. At higher concentrations between 0.25 and 0.50 g/dL, effects of alcohol range from impaired consciousness, sleep, stupor to unconsciousness, coma, and possibly death [97]. Hence, it is not surprising that alcohol has negative effects on victim escape probability. Numerous studies have also shown that individuals under the influence of alcohol do not respond to alarm systems, or respond at a decreased capacity, which may inhibit their ability for escape [20, 98, 99].

Since both alcohol and carbon monoxide target the central nervous system and act to depress it, their combined effects further inhibit occupant decision-making processes and escape ability. Mitchell et al. [100] found that rats failed to remain on a rotating rod or withdraw their feet from electric shock pads was substantially affected when alcohol was combined with CO exposure in comparison to either exposure alone.

4.2.7 Breaks in the Decision-making Process: Seeking Additional Information

Obtaining precise information about the situation during a fire emergency can prompt effective and safe action, reduce stress and support the problem-solving process. However, in a fire emergency, what is more likely is the receipt of ambiguous, incomplete and inconsistent information upon which decisions are made and actions are taken. If precise information is not obtained, occupants will likely engage in information-seeking actions or go back to their original pre-fire actions. This is important for engineers to understand because engaging in actions to seek additional information can cause significant time delays in taking protective action.

Cues from building signaling systems are ambiguous, and thus, occupants do not immediately begin evacuation

"The code alarm provisions generally assume fire alarm systems will enable building occupants to initiate effective emergency egress behaviors with minimal delay. Almost without exception, however, the technical literature contradicts this notion and supports the idea that an alarm signal alone does not generate immediate action – but starts the search for 'confirmation of a second clue" [5].

During unannounced evacuation drills in an underground station, passengers started their evacuation within 1 min when provided with precise voice communication instructions regarding the fire event, its location and action to be undertaken. With limited information such as the fire alarm bells only, many occupants were still in the station after 15 min [42].

In a fire, information seeking may involve consulting with others (milling), when others are accessible, or asking staff about the situation. The person may also initiate behavior to define the nature of the incident. This behavior often implies moving in the direction of the potential danger to gather information. These reactions to ambiguous information suppose that a certain amount of time is spent either ignoring the cues or investigating to define the situation.

Even in the most intense situations, people will seek information

Even after something as intense as a commercial airplane hitting World Trade Center 1 in the 2001 Disaster in New York City, the majority of people in both towers began milling immediately after WTC 1 was hit [101]. Many were interested in finding out what was going on or whether something was going on at all. Others were convinced that they knew what was taking place but were uncertain about what to do.

> Examples of information seeking are the WTC Tower occupants were looking around the floor, looking outside the window, and discussing the event with other people, both inside and outside of the towers. A person on the 40th floor of WTC 1, who originally thought that an earthquake had occurred, looked out the windows after the tower stopped shaking. Since his tower was still standing, he wondered about the other buildings and whether they were damaged (WTC1/040/0004). Similarly, a WTC occupant thought that underground steam pipes had exploded. He turned to look out the window and onto the concourse "trying to see if there [were]. . .emergency trucks or whatever, down there some place. . ." (WTC2/052/0003). Individuals also looked to one another for clues about what was going on. After hearing a loud sound, a WTC 2 occupant on the 84th floor scanned the trading floor, looking at others to see if they heard the sound as well (WTC2/084/0010).

The PADM accounts for the fact that individuals will seek information in an emergency event (See Fig. 4.2). If at any stage in the decision-making process the individual is uncertain (for example, during the comprehension stage, the person does not understand the information provided), the individual is then likely to engage in additional information-seeking actions. Information seeking is especially likely to occur when individuals think that time is available to gain additional insight on the question at hand. If information seeking is successful, in that the person at risk judges he/she has obtained enough information, then the individual moves on to the next stage in the decision-making process. However, if the information-seeking action is unsuccessful, there will be additional searching for information as long as the individual is optimistic that other sources or channels can help [49]. If the individual is pessimistic regarding future information-seeking success, the individual is likely to attempt to decide on a protective action based solely on whatever information is available.

The decision-making process is a circular one. Each time an individual receives new information, the decision-making process can begin again, either to alter or confirm original decisions made, if any. An individual who gains additional information is likely to carry on with the decision-making process until he/she is ready to implement a protective action.

Additionally, individuals do not have to go through each stage or question in the decision flow chart [49]. For example, if an individual is presented with information about the event from a credible source or if ordered to evacuate, the individual may move on to later stages in the decision process rather than going through each one in succession.

> **Steps in the decision-making process can be skipped, reducing delay time, if a credible source provides instructions on what to do in an emergency**
>
> In the 2001 WTC Disaster, Kuligowski [101] documented that some individuals decided that protective action was necessary for them, even without assessing a threat or risk to themselves or others. These individuals decided to evacuate strictly based on the message provider, even if they themselves did not believe the situation was dangerous. Instead, occupants simply followed instructions to evacuate because they were provided by someone with perceived authority or credibility. Many times, the source of the message was someone of higher status in the company, such as a manager. If their boss was telling them to evacuate, they were going to listen (WTC2/099/0001), even if they did not feel a sense of danger. An occupant in WTC 1 reflects on the day, saying that at the time he received instructions to evacuate from his boss:
>
>> . . .[t]here was no imminent reason to evacuate as far as we were concerned at this time. I mean there [was] no alarm. I don't think the alarm went off. I don't remember them going off. Otherwise there would be no reason to evacuate (WTC1/027/0002).
>
> However, he decided to evacuate because he was told to do so by his superior. In other examples, individuals followed instructions if they were given by occupants perceived as being knowledgeable about disasters, including fire wardens and 1993 survivors.

4.3 The Myth of Panic

A common expectation about human behavior in fire is the assumption that during a fire, occupants will panic. The possibility of panic behavior in a fire has been considered a "myth" by social scientists since the 1970s [102, 103, 104]. Although the media are very fond of this concept for its drama and sensational connotation, there is little evidence of panic in actual fire situations. It is a widespread misconception to believe that people caught in a fire will panic and try to flee in a stampede, crushing and fighting others. Panic supposes irrational behavior. On the contrary, people appear to apply rational, altruistic decision-making in relation to their understanding of the situation at the time of the fire. In retrospect, it is easy to point to some decisions that were not optimal and played a negative part on the outcome of a fire; however, at the time of the fire, these decisions are quite rational when all factors were considered.

It is commonly observed during interviews following fire events that victims themselves mention that they had panicked during the event. The public often uses the word "panic" as synonymous for being frightened, scared, nervous or anxious; usually it does not have the implication of irrational behavior. The limited knowledge that people have on fire development and fire dynamics does not prepare them for the best responses during fires. A majority of people who are faced with a fire situation react in a rational manner, considering the ambiguity of the initial cues, their limited knowledge about fires, and the restricted time they have to make a decision and to take action.

In the initial moments of a fire, upon smelling smoke or hearing the fire alarm, it is often observed that occupants do not react and deny or ignore the situation. This is labeled in the social sciences as normalcy bias [105]. This seems especially true in public buildings where occupants do not want to overreact to a false alarm or a situation that seems already under control. Such underestimation or acceptance of a potentially dangerous situation, often, results in delays in starting evacuation or in taking protective action. Once the situation has been defined as a fire, it does not necessarily imply that people will evacuate right away - time could be spent gathering family members and pets, getting dressed according to the weather, or finding keys, a wallet, and/or a purse.

4.4 Impact of Human Behavior in Fire on Fire Protection Engineering Design and Analysis

In summary, response to fires and other disasters is the result of a process. Building occupants or groups of occupants engage in a decision-making process (i.e., a series of steps) before they respond, based upon the cues presented from their environment. Before protective actions are taken, cues or information must be perceived (e.g., heard or seen), paid attention to, and comprehended before any actions take place. Therefore, it is important to understand that any information meant for building occupants must be provided in such a way to ensure that these three processes take place. Another important take-away is that occupants must perceive a credible threat and personalize the risk before taking action. In building fires, for example, occupants who witness heavy, thick, black smoke that decreases visibility and irritates the eyes are more likely than those noting less intense cues to realize that a serious event has taken place that puts them in danger [106].

As a consequence of this decision-making process, actions are performed. In instances where occupants do not feel at risk or that there is a threat to them, they are more likely to engage in normal, routine activities that delay them from taking protective action. Similarly, in circumstances where they perceive the threat or risk to them as uncertain or unclear, occupants are more likely to engage in information seeking activities that also delay their movement to safety. Research has shown that providing specific warning information in certain ways or providing leadership to prompt evacuation response could reduce the need for information seeking, and even the performance of certain protective actions, which allows occupants to reach safety sooner.

Understanding human behavior in fires can impact the fire protection engineering community in a number of ways; specifically, the ways in which we calculate evacuation timing for building designs, create and disseminate warning information, and develop methods for training building occupants to respond to fire events.

4.4.1 The Impact of Human Behavior on Evacuation Timing Calculations

Depending upon the circumstances, occupants can take a considerably long time before reaching a place of safety (within or outside of the building). Kuligowski et al. [107] found that, from observations of 14 building evacuations throughout the United States, the time that occupants take before entering the stairs can account for over 50% of the overall evacuation time. Therefore, practitioners should account for these delays in some way when calculating evacuation timing in a proposed building design fire. Actions and delay times associated with these actions can be especially important in certain types of buildings, where affected individuals are likely to engage in certain types of lengthy actions (i.e., those in which people may be asleep or located on upper floors of uniquely tall buildings).

Many times, when performing evacuation calculations, engineers are asked to provide a specific pre-evacuation time period or distribution as input. It is important to choose a time that is based upon specific scenarios expected and resulting occupant actions (and action timing).

4.4.2 Human Behavior Considerations Related to Warnings or Messages

Understanding of human behavior is important for the development of emergency messages and the dissemination of those messages before and during a fire event. The main way to prompt safe, effective, and appropriate action from building occupants is to disseminate warning messages during fire emergencies that will positively influence risk identification and assessment. Research has shown that a successful warning message contains the following factors or qualities:

- Specific about the threat and the risk involved [108–110],
- Repetitive [49],
- Consistent [111],
- Disseminated via multiple channels [112],
- Provided by a credible source [48, 113–115].

Source credibility is defined in terms of the source's expertise, including access to special skills or information, trustworthiness, or the perceived ability to communicate information about the disaster without bias [49, 116]. Source credibility can differ depending upon a number of factors, including the type of disaster, characteristics of the source, such as social role and believability, and characteristics of the warning receiver, such as past experience in disasters and social location [117–122]. For some warning receivers, credible sources may be friends and relatives, and for others, credible sources may be disaster authorities, such as government officials [123, 124] or fire fighters [50].

Additional information on emergency messaging can be found in Chap. 12.

4.4.3 Human Behavior Considerations Related to Occupant Emergency Training

Another way to prompt safe, effective, and appropriate action from building occupants is through training. An individual's past experiences in emergencies, specifically the actions that he/she has performed previously, can influence the actions that he/she considers as options during the current emergency [49, 125, 126]. The individual uses memories of the protective actions performed in the past as options for actions to perform in the current emergency. Similarly, an individual's emergency-based training and knowledge, for example, knowledge about evacuation procedures, can influence the options that one develops during an emergency [67, 127–130]. Guidance on training building populations can also be found in Chap. 12 [131].

4.5 Summary: Behavioral Facts

A great deal of information has been provided on human behavior in fire in this chapter. "Behavioral facts," first introduced by Kuligowski and Gwynne [132] and extended by Kuligowski et al. [133], are listed below to summarize the major findings captured by this chapter. A total of 28 behavioral facts are listed here:

- Previous experience of false alarms or frequent drills can reduce sensitivity to alarm signals, inhibiting perception processes.
- Some individuals exhibit hypervigilance that makes them particularly sensitive to certain cues.
- Habituation, focus and stress can narrow the perceptual field, and thus, not all available cues will be internalized.
- Sensory and cognitive impairments can inhibit the perception of cues.
- Content and clarity of the cue matters. The more clearly presented, without jargon, the more likely it will be comprehended accurately.
- The precision, credibility, consistency, comprehensiveness, intensity and specificity of the external cues will affect the assessment of the situation and perception of risk.
- Authority of the information source affects the perceived credibility of the information, and in turn the assessment of the situation and risk.
- Normalcy bias and optimism bias are commonplace. In other words, people often think that nothing serious is taking place, and that nothing bad will happen to them, respectively.
- Training on and/or experience with a particular incident type may allow a similar incident to be defined more quickly by the evacuee.
- The actions of the surrounding population can influence the internal processes of the individual.

4.5 Summary: Behavioral Facts

- People tend to satisfice rather than optimize.
- Pre-event commitment to a particular activity may cause individuals to decide against taking protective action.
- Authority of the source... affects the perceived credibility of the action.
- The actions of the surrounding population can influence the options of actions developed by the individual.
- Gender can influence the selection of actions to protect property.
- Social and authoritative roles, and social connections, can influence the selection of actions to help others.
- Training and experience in previous fire/evacuation events can influence the search for and selection of a particular action or set of actions.
- The appearance of a route can influence its use.
- The presence of smoke does not always preclude the use of a route, but can influence movement along a route.
- Training and experience may increase an individual's familiarity with the use of components/devices and subsequently improve their use.
- People have different abilities that influence actions taken.
- People seek information in situations where information is lacking or incomplete.
- People engage in protective actions, including preparing to move to safety or helping to protect others from harm, before they initiate a movement towards safety. These actions can also occur while moving to safety.
- People move towards the familiar, such as other people, places and things.
- People may re-enter a structure, especially if there is an emotional attachment to the structure, the contents and/or the inhabitants.
- People will behave in a rational AND altruistic manner; panic is rare.
- Evacuation is a social process, in that groups are likely to form during an evacuation.
- Social norms (or rules) in place prior to a fire event form the basis of those employed during the event.

Effects of Fire Effluent

5

5.1 Effects of Exposure to Smoke and Smoke Components

Once a victim has become trapped or incapacitated in a fire, conditions can become lethal within a few seconds to minutes; unsuppressed flaming fires can grow exponentially as does the heat, smoke, and toxic gases produced by these types of fires. The key determinant of survival in a fire is the time to incapacitation. Incapacitation occurs when the occupant is no longer capable of self-preservation due to exposure to toxicants, irritants, and heat in a fire. The acute physiological fire hazards affecting escape capability include:

- Impaired vision from smoke obscuration.
- Impaired vision, pain, and breathing difficulties from effects of smoke irritants on eyes and respiratory tract.
- Asphyxiation from toxic gases leading to confusion and loss of consciousness.
- Pain to exposed skin and respiratory tract followed by burns from exposure to radiant and convective heat leading to collapse.

The effects on escape capability tend to occur in this order and can be simply behavioral, when an occupant is deterred from entering a contaminated escape route by seeing smoke or flames, or behavioral and physiological when they attempt to move though fire effluent but their progress is inhibited by visual obscuration or eye irritation from smoke coupled with pain from heat. Severe physiological incapacitation can occur due to pain and breathing difficulties from exposure to irritant smoke, collapse from exposure to asphyxiant gases, or severe pain and burns from heat exposure.

The incapacitating effects of these hazards from smoke obscuration, irritants, asphyxiants and heat are summarized in the following sections. Calculation expressions are presented in Sect. 7 for assessment of incapacitation effects of each hazard. These expressions, the derivations for which are detailed in the *SFPE Handbook of Fire Protection Engineering* Chap. 63 [134], are designed to predict effects on exposed subjects engaged in a range of activities from sleeping to running. The default activity level is based on walking rapidly or descending a stair during escape activities. A somewhat simplified set of expressions is presented in the international Standard ISO 13571 (2012) which differ in some respects to those in the SFPE Handbook of Fire Protection Engineering [134]. The general derivation and results of applying these two approaches to an adult engaged at the default activity level are similar, but the following points may need to be considered:

- For smoke obscuration the author's proposed visibility tenability limit for escape impairment in [134] is 5 m (based on the potential for way finding difficulty and turning-back behaviour in irritant smoke) compared with 0.5 m in ISO 13571 (2012) (based on arm's length visibility). See Sect. 5.2 of this Guide for a detailed discussion on visibility in smoke.
- For irritant acid gases the proposed tenability limits for escape impairment are lower than for ISO 13571, but those for incapacitation are similar (see Sect. 7).
- For asphyxiant gases (CO, HCN, CO_2) the calculations give almost identical results for the default activity level. Reference [134] also includes the effects of oxygen depletion hypoxia. This is omitted from ISO 13571 on the basis that in most fire scenarios the combined effects of the other asphyxiant gases will exceed tenability limits before the contribution of the oxygen hypoxia term becomes significant.

© Society of Fire Protection Engineers 2019
SFPE Society of Fire Protection Engineers, *SFPE Guide to Human Behavior in Fire*, https://doi.org/10.1007/978-3-319-94697-9_5

- The expressions for the effects of exposure to hot environments (convected heat) are the same in both, but the expression for effects of radiant heat are somewhat different. Reference [134] has a slightly more conservative expression for time to pain from heat radiation, but also includes expressions for time to secondary and full thickness burns. It also discounts the effects of clothing, on the basis that hands and head are likely to be unprotected.

5.1.1 Asphyxiants

Asphyxiant gases cause incapacitation through effects on the central nervous and cardiovascular systems. Most asphyxiant gases produce their effects by causing brain tissue hypoxia i.e. reducing the amount of oxygen delivered to the brain. Since the body possesses powerful adaptive mechanisms designed to maximize oxygen delivery to the brain, it is possible to maintain normal body function up to a certain dose of asphyxiant. The victim is often unaware of their impending intoxication, and once normal function can no longer be maintained, deterioration is rapid and severe. Signs of asphyxiant exposure may include lethargy and euphoria with poor physical coordination followed by rapid unconsciousness and death.

The two major asphyxiant gases in fires are carbon monoxide (CO) and hydrogen cyanide (HCN). Carbon monoxide is always present in some concentration in all fires, irrespective of the materials involved or the stage (or type) of fire. Hydrogen cyanide is produced when nitrogen-containing materials are burning in the fire. These include materials such as polyacryonitriles, polyurethane foams, melamine, nylon, and wool, which are fairly ubiquitous in modern homes.

5.1.1.1 Carbon Monoxide

Carbon monoxide combines with hemoglobin in the blood to form carboxyhemoglobin (COHb), which results in a toxic asphyxia by reaching the amount of oxygen supplied to the tissues of the body particularly in the brain, Table 5.1 [134] provides descriptions of the physiological effects which result as blood COHb levels increase in the blood.

At COHb concentration below 10%, symptoms of exposure are minimal with a slight reduction in exercise tolerance. Between 10 and 20%, occupants will begin to experience headaches and abnormal vision. As concentrations increase, symptoms become more severe; between 20 and 30% COHb, individuals may experience headaches, nausea, and loss of fine motor skills progressing to incapacitation at ranges of 30–40% COHb. Once an occupant reaches a 50% COHb concentration in the blood, death is likely imminent. In the case of individuals with compromised cardiac conditions, levels as low as 20%

Table 5.1 Classical Relationship between Carboxyhemoglobin Concentration and Signs Exhibited in Humans and Non-Human Primates

Blood saturation %COHb	After Stewart [135]	After Sayers and Davenport [136]	After Purser [137–139] (non-human primates)
0.3–0.7	Normal range due to endogenous production		
1–5	Increase in cardiac output to compensate for reduction in oxygen carrying capacity of blood (heart patient may lack sufficient cardiac reserve)		
5–9	Exercise tolerance reduced, visual light threshold increased, less exercise required to induce chest pains in angina patients	Minimal symptoms <10%	
16–20	Headache, abnormal visual evoked response, may be lethal for patients with compromised cardiac function	Tightness across forehead and headache experienced 10–20%	
20–30	Throbbing headache; nausea; abnormal fine manual dexterity	Throbbing headache	
30–40	Severe headache; nausea and vomiting; syncope (fainting)	Severe headache; generalized weakness, visual changes; dizziness, nausea, vomiting, and ultimate collapse	30% caused confusion, collapse and coma in active animals during 30-min exposures with nausea after
40–50		Syncope, tachycardia (rapid heartbeat) and tachypnea (rapid breathing)	40% caused coma, bradycardia (slow heartbeat), arrhythmias, EEG changes, in resting animals during 30-min exposures
50+	Coma and convulsions	Coma and convulsions	
60–70	Lethal if not treated	Death from cardiac depression and respiratory failure	

5.1 Effects of Exposure to Smoke and Smoke Components

COHb have proven to be lethal, so design evaluations should give consideration to the type of occupant being exposed. Carbon monoxide accumulation in the blood is dose dependent, e.g. the concentration achieved in the blood is not only a function of the inhaled carbon monoxide concentration in the air but also a function of the duration of exposure to the toxicant (see Sect. 7.1.1.1).

5.1.1.2 Hydrogen Cyanide

Hydrogen cyanide is a toxic gas with acidic properties, causing toxic asphyxia when inhaled. The pattern of incapacitation for HCN is somewhat different from that produced by CO in that the effects occur more rapidly. Unlike CO, HCN is not held primary in the blood but is carried rapidly to the brain. Although the accumulation of a dose is one factor, the most important determinant of HCN incapacitation appears to be the rate of uptake, which in turn depends on the HCN concentration in the smoke and the subjects' respiration. Table 5.2 provides a summary of effects in humans and primates that result from the specified exposure doses (concentration x time), which vary with the exposure concentration. For example, studies conducted by Purser [79] showed that primates and humans exposed up to 80 ppm of HCN for up to an hour experienced only mild hyperventilation (3600 ppm. minutes), while exposure to 200 ppm resulted in incapacitation after two minutes (400 ppm. minutes).

The data in Table 5.2 suggests that the effect of exposure to HCN varies significantly between 200 ppm and 530 ppm. Kimmerle [140] reported that death occurs "rapidly" in humans at concentration of HCN greater than 300 ppm, however, Barcroft [142] found that a man survived a 15-min exposure to 530 ppm of HCN. With HCN and CO, body size does appear to influence time to incapacitation, therefore, it would be expected that a human would tolerate exposure to a given concentration longer than a cynomolgus monkey; this same concept is true for adults in comparison to children. Additionally, physical activity causes more rapid uptake due to increased respirations so an actively escaping subject will be overcome approximately three times more rapidly than a sedentary subject.

5.1.2 Hypoxia

Apart from the tissue hypoxia caused by CO and HCN, hypoxia in fires can also be caused by exposure to low oxygen concentrations. The body has compensatory mechanisms that maximize the supply of oxygen to vital organs when low oxygen concentration is encountered in the inspired air (e.g. high altitude) or in the lungs (e.g. during exercise). However, these compensatory mechanisms can become overwhelmed when oxygen depletion becomes significant. Four hypoxic phases have been identified: [134] indifference, compensation, manifestation, and critical with corresponding oxygen concentration ranges of 14–21%, 12–14%, 10–12%, and 8–10%, respectively. Little to no effects on physiological function is seen during the indifference phase with only slight increases in ventilation and heart rate during the compensation phase. However, once the manifestation phase is reached, degradation of higher mental processes and neuromuscular control, loss of critical judgment and volition with dulling of the senses, and a marked increase in cardiovascular and respiratory activity will result. Critical hypoxia is typically reached once the oxygen concentration drops below 10%. This concentration marks the point at which rapid deterioration in judgment and comprehension occurs and leads to unconsciousness. Hence, it should be assumed that escape would not be possible at oxygen concentrations below 10%. Table 5.3 provides a summary of exposure effects to low oxygen concentrations.

Table 5.2 Reported Effects of HCN Inhalation

HCN (ppm)	Species and effects	Author
<80	Mild background hyperventilation with exposures up to 1 h in primates and humans	Purser [79]
100	Loss of consciousness after 23–30 min in primates and humans	Purser, Kimmerle [79, 140]
200	Loss of consciousness after approximately 2 min in primates and humans	Purser, Kimmerle [79, 140]
300+	Death occurs "rapidly" in humans	Kimmerle [140]
444	A man survived an accidental exposure	Bonsall [141]
530	A man survived a 1.5-min exposure without collapse – his dog exposed at the same time was comatose and almost died	Barcroft [142]
539	Suggested 10-min LC_{50} in humans	McNamara [143]
1000	One breath may cause loss of consciousness in humans	Purser [79]

Table 5.3 Exposure effects to low oxygen concentrations

Oxygen %	Hypoxic phase	Effects
14–21	Indifference	Little/no effect on physiological function
12–14	Compensation	Slight increase in ventilation/heart rate
10–12	Manifestation	Degradation of higher mental processes and neuromuscular control; Loss of critical judgment and volition; Dulling of the senses; Marked increase in cardiovascular and respiratory activity
8–10	Critical	Rapid deterioration in judgment and comprehension leading to unconsciousness: Assumed that escape would not be possible at oxygen concentrations below 10%

5.1.3 Carbon Dioxide

Carbon dioxide (CO_2), like carbon monoxide, is universally present in fires, but typically at low concentrations: on the order of 0.4%. Although carbon dioxide is not toxic at concentrations of up to 5%, it stimulates breathing. At 3% the respiratory rate is approximately doubled, and at 5% the respiratory rate is approximately tripled. This hyperventilation, apart from being stressful, can increase the rate at which other toxic fire products (such as CO and HCN) are taken up. Therefore, it is likely that the increased uptake of asphyxiants resulting from carbon dioxide induced hyperventilation will significantly reduce time to incapacitation. Hence, this is an important factor which needs to be accounted for in a hazard analysis.

5.1.4 Irritants

Irritant fire products cause incapacitation during and after exposure in two distinct ways. During exposure the most important form of incapacitation is sensory irritation, which causes painful effects to the eyes and upper respiratory tract and to some extent the lungs. Although irritant exposure may be painful and thus incapacitating, it is unlikely to be directly lethal unless exceptionally high concentrations of irritants are present. However, the second effect of irritant exposure is an acute pulmonary irritant response, consisting of edema and inflammation. This pulmonary response can cause respiratory difficulties and may lead to death within 6–24 h after exposure [144, 145].

For sensory irritation the effects do not depend on an accumulated dose but occur immediately on exposure and usually lessen somewhat if exposure continues [146, 147]. For the later inflammatory reaction, the effect does depend on an accumulated dose; there appears to be a threshold below which the consequences are minor, but when this dose is exceeded, severe respiratory difficulties and often death occur [134].

The main effects of sensory irritants are to enhance the effects of smoke particulates on distress, visual impairment, wayfinding, willingness to attempt to enter, walking speed in escape routes, plus the additional incapacitating effects of eye and respiratory tract pain, and breathing difficulties. After rescue the main hazard of death at the hospital is from lung edema, inflammation, and pneumonia.

5.1.5 Toxic Fire Gas Interactions

Data on interactions between CO, HCN, NOx, irritants, low oxygen, and CO_2 are limited, but where deleterious interactions are likely, it is prudent to include them in the incapacitation model. It is proposed that the interactions should be quantified in the incapacitation model as follows:

1. CO and HCN are directly additive (1:1) on a fractional dose basis.
2. The asphyxiant effects of NO and NO_2 (designated as NOx in mixtures) are additive with those of HCN and CO.
3. Irritants will cause some level of hypoxia (additive term is included in the FLDirr equation).
4. For design purposes, inputs for an adult engaged in light work such as walking along an escape route should be utilized. The rate of uptake of asphyxiant gases (CO, HCN, NOx and irritants) depends on the respiratory ventilation (VE) of the subject in relation to body size, which in turn depends upon their level of physical activity.
5. The main effect of carbon dioxide is to increase the breathing rate and thus the rate of uptake of asphyxiant gases (CO, HCN, NOx and irritants). A multiplicatory term VCO_2 should be used when calculating this effect.

5.1 Effects of Exposure to Smoke and Smoke Components

6. Low oxygen hypoxia is additive with the overall hypoxic effects of CO and HCN.
7. The beneficial effects of increased CO_2 on the hypoxic effects of CO and low oxygen hypoxia are ignored.
8. The direct intoxicating effects of CO_2 are considered unlikely to occur before other effects, and therefore, are normally ignored (unless exposure to 7% CO_2 or higher is anticipated).

Taking into account these assumptions, it is possible to derive a general expression for calculating uptake, fractional effective doses, and time to incapacitation from asphyxia as described in the analysis section.

5.1.6 Heat

There are three basic ways in which exposure to heat may lead to incapacitation and death: heat stroke, body surface burns, and respiratory tract burns [134].

5.1.6.1 Heat Stroke (Hyperthermia)

Simple hyperthermia involves prolonged exposure (approximately 15 min or more) to heated environments at ambient temperatures too low to cause burns. Under such conditions, where the air temperature is less than approximately 120 °C for dry air or 80 °C for saturated air, the main effect is a gradual increase in the body core temperature. Increases above the normal core temperature of 37 °C, up to approximately 39 °C, are within the physiological range, and can occur at normal ambient temperatures during hard exercise. However, once 40 °C is reached, consciousness becomes blurred, and the subject becomes seriously ill. Further increase causes irreversible damage with temperatures above 42.5 °C being fatal unless treated within minutes. Table 5.4 provides a list of tolerance times at various temperature conditions.

5.1.6.2 Skin Burns

According to Buettner [152], pain from the application of heat to the skin occurs when the skin temperature at a depth of 0.1 mm reaches 44.8 °C. This is in agreement with the research of Lawrence and Bull which found that discomfort was experienced when the interface between a hot handle and the skin of the hand reached 43 °C. The sensation of pain is followed soon afterward by burns, causing incapacitation, severe injury, or death, depending on their severity. The time from the application of heat to the sensation of pain and from pain to the occurrence of burns of various degrees of severity depends on the temperature, or more properly, the heat flux to which the skin is exposed. The effects of heating the skin are essentially the same whether the heat is supplied by conduction from a hot body, convection from air contact, or by direct radiation. Pain, therefore, occurs when the rate of supply of heat to the skin surface exceeds the rate at which heat is conducted away by an amount sufficient to raise the skin temperature to 44.8 °C.

5.1.6.3 Thermal Damage to the Respiratory Tract

Heat damage to the respiratory tract is even more dependent on the humidity of inhaled hot gases than are skin burns. As a result of the low thermal capacity of dry air and the large surface area of the airways (which are lined with a wet surface and good blood supply) thermal burns are not induced by dry air below the top of the trachea. However, steam at around 100 °C is capable of causing severe burns to the entire respiratory tract down to the deep lung due to its higher thermal capacity and the latent heat released during condensation. These effects of inhaled hot gases are demonstrated by the work of Moritz et al. [153, 154] in which anesthetized dogs and pigs breathed hot air, flame from a burner, or steam, supplied through a cannula to the larynx. Dry air at 350 °C and flame from a blast burner at 500 °C caused damage to the larynx and trachea but had no effect

Table 5.4 Reported tolerance times for exposures to hot air

Temperature (°C)	Time (min)	References
Dry air		
110	25	Simms and Hinkley [148]
180	3	Simms and Hinkley [148]
205	4 Bare headed, protected	Veghte [149]
126	7	Elneil [150]
Humid air		
32 at 100% RH	32 Men working	Leithead and Lind [151]

on the lung, whereas steam at 100 °C caused burns at all levels. Hence, heat flux and temperature tenability limits designed to protect victims from incapacitation by skin burns should be adequate to protect them from burns to the respiratory tract as well. In practice, 60 °C has been found to be the highest temperature at which 100% water-vapor saturated air can be breathed [134].

5.2 Visibility/Smoke Obscuration

The subject of visibility is an important consideration in the tenability analysis and a number of independent researchers and scientists have put forth their own individual suggestions for visibility limits and/or associated extinction coefficient values/ optical density values. These individual's suggestions [155–159] range from visibility limits of 1.2 m to as high as 20 m including tests with illuminated exit signage, normal lighting, and irritant and nonirritant smoke. It is recognized that, for a wide variety of reasons including personal judgments, any one of these values could be selected as a criterion to use in engineering analysis. However, simply selecting a fixed visibility threshold (e.g. 13 m) does not provide engineering certainty that an individual can or cannot escape or evacuate through smoke. Milke et al. [160] has noted that visibility should be treated very differently than an exposure to gas concentrations or heated smoke. There are key tenets to consider when addressing visibility.

- The reduction of visibility along a path of egress will not by itself incapacitate individuals or cause death. Excessive thermal exposure and high doses of toxic gases produced by a fire can result in incapacitation and death.
- Lack of visibility (short term) does not incur a physiological effect, meaning people are capable of traveling through dark rooms. Traveling can be expected to be slowed by darkness but direct injury is not attributable to darkness (short term) alone. Similarly, people have been shown to move through dense smoke at varying speeds in experiments and in some fire incidents.
- Whether or not occupants will enter and move through smoke during fire incidents depends on a number of aspects of the subjects and the fire scenario. Subjects may attempt to move through dense smoke when they find themselves in a fire or smoke-filled enclosure. In such scenarios, they are highly motivated to move through smoke in an attempt to escape. However, when occupants are in a relatively clear environment with the option of remaining or turning back to a place of refuge, they are forced to make a judgment on whether to enter a smoke-filled escape route or take refuge and await rescue.
- Reduced visibility results in reduced movement speed and longer exposure to heat and toxic gases. With sufficient, continued exposure to heat and/or toxic gases speed of movement can continue to slow occupant movement and diminish occupants' mental capabilities. Reduced visibility also affects wayfinding and may increase the time taken to locate exits. If evacuation is not completed, incapacitation may result.
- People will move through smoke. Bryan [30] cites several studies that illustrate that occupants will move through smoke to evacuate a fire situation and that such movement may often occur in limited visibility conditions (3–4 m) and perceived worsening smoke conditions. However, it is also noted that some percentage of occupants will not complete the evacuation and turn back in reduced visibility conditions. In these studies, approximately 30% were found to turn back rather than continue through smoke-logged areas. The average density at which people turned back was at a "visibility" distance of 3 m. This represents an optical density (OD·m^{-1}) of 0.33, (extinction coefficient 0.76 (1/m)) with women more likely to turn back than men. A difficulty with this kind of statistic is that, in many fires in buildings, there is a choice between passing through smoke to an exit or turning back to take refuge in a place of relative safety, such as a closed room. In some situations, people have moved through very dense smoke when the fire was behind them, whereas in other cases people have failed to move at all.
- Behavior might also depend on whether smoke layering permits occupants to crouch down to levels where the smoke density is lower and visibility permits continued wayfinding.
- An individual's ability to progress through smoke depends on both the optical density and the irritancy of the smoke. Based on the finding that people move as if in darkness at a visibility of 5 m in irritant smoke and that smoke from most fires contains a variety of irritant chemical species, generic design tenability limits have been proposed on the basis that concentrations exceeding these levels could impair or even prevent occupants' safe escape (for situations where the concentrations of acid gases are considered unlikely to be significant due to the fuel composition) (See Table 7.8) [134].
- It is noted that in those studies citing "turn-back" behavior, the reduced visibility conditions did not result in fatalities as all occupants were able to participate in the after-fire surveys, so returning to take refuge is sometimes a successful survival strategy. Whether or not occupants survive having turned back depends upon the subsequent fire development. Fire

incident investigations have revealed the following scenarios: (1) cases where occupants have survived, because they escaped through dense smoke, (2) cases where they have died attempting to escape through smoke when they could have survived had they sheltered in place, (3) cases where occupants have survived because they sheltered in place, and (4) cases where they have died after they remained in place, but arguably could have escaped through smoke and survived. These individual and scenario-related issues present a challenge in establishing a consensus value of visibility distance (as a deterministic endpoint criteria). A value of 0.5 m (1.6 ft) visibility is provided in ISO 13571:2012 for visual obscuration criteria as a condition which indicates occupants' tenability may be compromised. (ISO 13571:2012, "Life Threatening Components of Fires — Guidelines on the Estimation of Time Available for Escape Using Fire Data)." This consensus document prescribes an endpoint dependent upon the visibility reduction at which occupants are no longer able to take effective action to accomplish their own escape. The ISO 13571:2012 surrogate for this, in visibility terms, is when occupants can see only an arm's length distance of 0.5 m (1.6 ft). The assumption is that when such loss of visibility or obscuration occurs, occupants' escape will slow sufficiently and toxic gases or heat exposure can result in incapacitation. Again, it is important to note that such reduced visibility alone (0.5 m/1.6 ft) is not a fatal or incapacitating condition. True incapacitation occurs when the heat or toxic gas tenability criterion is exceeded.

Based on the above, it is clear that a reduction in visibility alone may not prevent escape and in itself does not constitute a visibility endpoint (tenability endpoint); it must be coupled with the potent, harmful effects of exposure to heat and toxic gases in cases where visual smoke obscuration inhibits or prevents timely escape before conditions become untenable. A primary consideration is, does the inhibition of evacuation, impairment of wayfinding, or reduction in movement speed due to reduced visibility sufficiently lengthen the evacuation and exposure process to a point that incapacitation occurs due to thermal and/or toxic gas exposure or result in occupants becoming trapped in a location where conditions subsequently become lethal due to fire spread or structural collapse. At present, most of the research concerning visibility of human subjects and its relationship to the ability to safely egress a structure has focused primarily on commercial and public buildings. Some research has been conducted in a residential setting, where occupants are typically very familiar with their surroundings, which shows occupants will in some cases cease to move or abandon efforts to escape when not incapacitated [134]. As such, using any of the above suggested visibility values or associated extinction coefficients as a performance endpoint for tenability can only be effectively evaluated in conjunction with an appropriate analysis of the occupant and fire scenarios with occupant gas and heat exposure, since reduced visibility alone does not cause incapacitation/fatalities.

Part II
Modelling Human Behavior in Fire

Development and Selection of Occupant Behavioral Scenarios

6

6.1 Introduction

SFPE's *Engineering Guide to Performance-Based Fire Protection* discusses the development of design fire scenarios for the evaluation of performance-based designs, but it only briefly touches on how occupant considerations factor into these assessments [3]. This chapter will provide more detail on the steps engineers should take and factors they should consider in demonstrating the adequacy of life safety provided by their designs, by developing occupant behavioral scenarios that complement design fire scenarios in order to explore the effects of different behavior within a given fire scenario.

Occupant scenarios include more than just the occupant characteristics discussed in Chap. 3. These scenarios are built around the interaction between the building characteristics, the fire environment, *and* the occupants.

6.2 Background

The simple model for the evaluation of an engineered design compares the time available for evacuation (Available Safe Egress Time, or ASET) with the time required to evacuate the occupants (Required Safe Egress Time, or RSET). When ASET is greater than RSET, with some not-yet-defined safety factor or safety margin applied, the engineered design is considered 'safe.'

ASET is generally calculated by fire models, based on design fire scenarios. RSET is generally calculated by egress models and movement calculations. The occupant behavioral scenarios described in this chapter are used to configure the approaches used to quantify RSET as well as to develop occupant movement strategies, as described in Chap. 13.

This method of evaluation uses fire hazard calculations to estimate the development and growth of a fire and the spread of resulting combustion products in order to estimate the time before untenable conditions exist in occupied spaces. Those results are then compared to the estimates from an egress calculation that will predict whether people will still occupy or pass through those untenable spaces and be exposed to critical conditions. Similarly, the combination of a fire model and egress model can be used to calculate Fractional Effective Dose (FED) and Fractional Effective Concentration (FEC) to see if individual occupants are predicted to experience exposure levels approaching incapacitation or lethality as they evacuate the building.

The use of a fire model in such an evaluation requires the development of appropriate design fire scenarios. The process of selecting design fire scenarios to test the safety of a building should be well-understood by fire protection engineers and will not be discussed in detail here. For details on the process of selecting design fire scenarios, see SFPE's *Engineering Guide to Performance-Based Fire Protection* [3]. Some regulatory codes or other guidance documents specify fire scenarios to be analyzed, including fires that block exits, concealed fires, smoldering fires, fires that threaten occupied areas, etc. [161, 162]. These fire scenarios in several cases mention or imply the presence of building occupants, but no details are explicitly specified as to the number or types of occupants who should be considered. We know from research in building fires that the ability of people to safely evacuate or relocate in case of a fire depends on their capabilities to recognize and react to warning cues, move or be moved to a safe location, choose exit routes, and make other decisions. When an engineer makes evacuation calculations, either by hand or with a computer model, little guidance is available.

© Society of Fire Protection Engineers 2019
SFPE Society of Fire Protection Engineers, *SFPE Guide to Human Behavior in Fire*, https://doi.org/10.1007/978-3-319-94697-9_6

This chapter will present a method for considering specifics about building occupants when predicting evacuation timing. To complement the development of design fire scenarios, this method considers the selection of occupant behavioral scenarios. The ultimate objective is to quantify these occupant characteristics in order to model the evacuation behavior.

6.3 Occupant Behavioral Scenarios

An occupant behavioral scenario can be considered a qualitative description of the characteristics, actions, and decisions of the occupants that should be included in a deterministic analysis. So it involves more than just the characteristics of the occupants that affect response and reaction (such as age, gender, mobility, and cognitive abilities, etc.) but also the choices they can make, many of which will be influenced by the building design (such as exit route choice). Many of these characteristics were described in Chapter 3. The engineer or designer will have to quantify the description of the occupants in the scenario, determining the number of occupants who will be present, assigning characteristics to those occupants, such as age and capabilities, and determining rules that they will follow, such as decision making and routing choices.

There are various inputs to the development of these occupant behavioral scenarios. These scenarios may be derived from this document or elsewhere. It is important that these inputs are appropriate for the specific scenario being developed. For instance, it would not be appropriate to automatically employ assumptions from prescriptive codes in an uninformed manner. This section describes the informed development of occupant behavioral scenarios.

6.3.1 Some Aspects of Occupant Scenarios are tied to the Fire Scenario

Three of the fire scenarios listed in the NFPA 101®*Life Safety Code*® [162] consider the impact of a fire on the occupants of the building. In two of the three cases, the fires are to start in a room or concealed space endangering or adjacent to a large room occupied by a large number of people, and the third specifies a fire close to a high-occupancy area. Another scenario places the fire in the primary means of egress. The consideration of the occupant population, then, really begins in the selection of design fire scenarios, but no guidance is given there as to the composition of the occupant population that will have to be specified for evacuation modeling when doing a full fire safety evaluation of the design.

The International Organization for Standardization (ISO) has developed a standard for the selection of design fire scenarios, [163] and recently completed a technical specification that provides guidance for the selection of occupant behavioral scenarios. This latter document demonstrates how occupant scenarios are tied to the fire scenarios that are relevant in evaluating a particular engineered design, [164] when the fire safety objective of the design analysis is life safety. That process will be the basis for the following discussion.

Each of the first six steps in that process for the selection of fire scenarios should require some thought as to the characteristics and capabilities of potential occupants of the building who will be impacted by the design fire scenarios:

- Identify the specific fire safety challenges
- Location of fire
- Type of fire (fire characteristics)
- Potential complicating hazards leading to other fire scenarios
- Systems and features impacting fire
- Occupant actions impacting the fire

The first step in selecting fire scenarios is to consider all the potential uses of the building under consideration, and for each use think about all the different types of users that would involve. For example, a hotel complex probably includes sleeping rooms upstairs, with function rooms, restaurants, shops, a fitness center, and/or a business center located in various locations throughout the building. The use of the sleeping rooms is fairly obvious. But the physical and cognitive abilities of the occupants of those sleeping rooms may vary. For example, some may be intoxicated. There might be children or elderly guests who will need special assistance. There will be some proportion of guests, possibly quite a large proportion, who have physical disabilities that will impact their ability to receive warning cues and act on them.

And how might the function rooms be used? They might be used for meetings, with a relatively small group of people seated around a large table. On other occasions, the same room might be filled with rows of seats to accommodate the maximum occupant load in the room for a seminar. It might be used for a party or dinner dance, with large numbers of people crowded around tables, leaving little circulation space. Who is using those rooms, and what are their expected abilities and capabilities? One might assume that those attending a meeting are awake, alert and sober. But they could have physical disabilities that would complicate their evacuation. Those attending a party, on the other hand, might be intoxicated, or they might include a large number of children who would require assistance. And it is necessary to anticipate that some percentage, and possibly a quite large percentage, might have sensory, cognitive or physical disabilities that would impact their ability to hear and understand warning systems and/or evacuate.

The second step involves determining the location of the fire. As demonstrated in the *Life Safety Code*[R]'s required scenarios, the focus here is on fires that are most likely to endanger the occupants of the building. There is no need to consider any fire that will not begin or spread flames and smoke to spaces that will not be occupied or used by the occupants.

The third step focuses on the type or characteristics of the fire. The type of fire will influence the time to activation of alarm systems, influencing the amount of time available for escape; flame and smoke spread into exit routes which can impede the evacuation or result in entrapment; and reduce visibility which will slow walking speeds and possibly obscure exit signs and visible messaging systems. Reduced walking speeds could result in longer exposure to a toxic atmosphere, resulting in incapacitation or death. A rapidly developing fire could cut off escape routes early in an evacuation.

The fourth step calls for consideration of complicating hazards, such as earthquakes which can cause fires and also disrupt travel routes (e.g., displacing stairwells or jamming doors) or introduce debris into travel paths, or power failures, that can reduce available lighting and could negatively impact signage and messaging systems. Although there is no similar concept in the *Life Safety Code*[R]'s specified scenarios, these potential challenges to a building's design are worth some consideration.

The fifth step looks at active and passive systems and features that could impact a fire. As far as occupant considerations are concerned, one of the issues to think about are occupant actions before a fire which can compromise the effectiveness of these systems. For example, will there be a fire safety management plan that will reduce the risk of building occupants propping open fire doors or doors into pressurized stairwells, or introducing openings in fire walls? In the normal use of the building, will interior doors be open or closed (thus affecting the spread of fire and combustion products)? In areas remote from the fire, a hot smoky layer may develop above a lower layer that poses little threat to evacuating occupants, but the activation of a sprinkler system could result in mixing of the layers, thus reducing visibility in exit paths.

The sixth and final step in developing fire scenarios takes a look at occupant actions after ignition that affect the growth and development of the fire. Some positive influences here would be trained staff who could enhance notification of other occupants and facilitate the evacuation, or an in-house fire brigade that could undertake suppression activities. Less positive occupant actions would include spreading debris in escape paths by abandoning possessions during an evacuation, creating tripping hazards and possibly increasing the fuel load along egress paths; or opening too many doors into a pressurized stairwell and reducing its effectiveness.

6.3.2 Identifying Occupant Scenarios

Now that the fire scenarios have been selected, the engineer or designer must determine with more specificity the occupant scenarios to be associated with each fire scenario. As a sample process, the following steps are outlined in the ISO technical specification for the selection of occupant behavioral scenarios.

In the first step, the number of occupants and their locations are determined (often this is determined as the maximum likely number which should never be exceeded). The process of deciding the location of the fire for the fire scenario will probably also determine the location of the occupants. In the *Life Safety Code*[R] for example, fires located near or adjacent to a large group of people are to be considered. Once the fire location is decided, that high-occupancy area is known. The number of occupants can be based on design considerations (the number of people it is meant to hold, for example, in a theater with fixed seating), or maximum densities specified in national codes can be used. Crowdedness is a well-recognized phenomenon and potential danger in an evacuation, but it is important to keep in mind that low densities can also provide challenges to occupants, in cases for example where there are fewer people to help spread a warning or share information.

In the next step, the characteristics of the occupant population are determined, in terms of their physical and cognitive capabilities. The occupant population should be chosen to provide an appropriate challenge to the design's safety, while being realistic. For example, a movie theater can hold a large number of people, but assuming that all the people in the theater are mobile, healthy and sober will not challenge the design sufficiently. That theater will at times be occupied by possibly large numbers of older adults with a range of mobility issues, or children who will need assistance. Occupants could be intoxicated and not capable of making good decisions on exit route choice. The familiarity of occupants to the location can also be a factor in their evacuation as will their vulnerability to the products of combustion from the fire.

Table 6.1 Example Occupant Characteristics Matrix

Characteristic	Hotel Guests	Restaurant patrons	Hotel employees
Familiarity	Transitory	Transitory	Permanent
Training	None	None	Yes
Ages	Adults and children	Adults; children possible	Adults
Disabilities?	Wide range possible	Wide range possible	Small range possible
Vulnerabilities	Possible	Possible	Possible
Level of intoxication	Intoxication possible	Intoxication possible	Conscious
Awake	Awake or asleep	Awake	Awake
Social groupings	Individuals, couples, families	Individuals, couples, families, groups	Individuals, co-workers
Role	Guest (expects assistance)	Guest (expects assistance)	Manager/ subordinate

The third step addresses the activities of the occupants. This will include whether they are awake or asleep, obviously, as well as the degree to which they are invested in their activities. For example, shoppers who are in line to pay for their purchases may be more reluctant to abandon their shopping and leave a store than those who have just walked into the store and may be more willing to turn and leave. In some cases, people can be so engrossed in their activity that they fail to recognize and respond to cues about the fire.

And finally, the presence of any staff trained in evacuation should be considered. Trained staff can positively affect the occupants' response to alarms by encouraging people to suspend their activities and begin to evacuate, thus reducing their pre-evacuation time, as well as directing people to the nearest exit, which can reduce their travel time to safety.

Having taken into consideration the potential uses and users of the planned building and having factored in all the actions and characteristics of the users that can impact their safety in the event of a fire, the engineer or designer can then construct a matrix of occupant characteristics from which to distill the smaller set of design occupant behaviors (analogous to design fire scenarios) that will be modeled in the evaluation.

Table 6.1 is from the fifth edition of the *SFPE Handbook of Fire Protection Engineering* [2] and shows a matrix of occupant characteristics that could be relevant in looking at a hotel design. In this example, the process of selecting design fire scenarios focused on three uses of the hotels – the sleeping rooms upstairs, the restaurant on the ground floor and the building as a workplace for hotel employees. The wide range of possible characteristics of the hotel guests is considered – they may be of any age, asleep during the night, with a range of abilities and disabilities, and with some possibly intoxicated. They would generally be unfamiliar with their environment. Restaurant patrons would exhibit the same potential range of characteristics and capabilities and also be generally unfamiliar with their environment. The staff on the other hand would be more familiar with the hotel and its systems, probably less likely to include a large proportion of people with disabilities and be less likely to be intoxicated or asleep.

When the fire scenarios are chosen, these are the types of people who must be shown to be safe in the evaluation of a proposed engineered design. The actual occupant scenarios that will be considered in evaluating the design will be selected based on available data or engineering judgment. The scenarios should be worst credible cases, and various options will need to be compared to each other in order to choose a selected set that sufficiently challenges the design. For example, the engineer or designer can determine that assuming a large number of elderly occupants in a building, with a range of sensory impairments that may delay their response time to an alarm and mobility impairments that will slow their walking speeds will challenge the design as much or more than many other scenarios with various combinations of mobility and sensory impaired occupants. That scenario, then, will be considered representative of a cluster of other combinations of occupant characteristics that could potentially be present.

6.4 Documentation

The choice of occupant behavioral scenarios must be documented to show the reasoning behind the final selection. The consequence and likelihood of various scenarios will have to be considered (See the *SFPE Engineering Guide to Fire Risk Assessment* [165] for additional information). While it may be reasonable to eliminate very unlikely scenarios, even when the potential consequence of any of those scenarios could be quite high, it is important to ensure that the elimination of many low-likelihood, high-consequence scenarios has not resulted in the elimination of a significant proportion of likely scenarios.

If a scenario is eliminated because no design could prevent death or injury, that must be documented. For example, it may not be possible to protect someone who is close to the initial fire location or to protect a building from a terrorist attack or multiple simultaneous deliberately-set fires. If, however, occupants could reasonably be expected to be vulnerable under certain conditions, such as patients in a hospital or intoxicated people in a club, protections *should* be included in the design to protect against such predictable threats.

If a scenario is eliminated because any solution would be cost-prohibitive, that must be documented, and approved by all stakeholders.

6.5 Quantifying Occupant Behavioral Scenarios for the Evaluation

When using an evacuation model to assess RSET for the proposed design and the chosen scenarios, the characteristics and elements of the occupant scenarios must be converted in some way to model inputs. Although evacuation models vary, there are five descriptors that most will most require be specified [166]. These descriptors are:

1. The delay time before occupants begin to evacuate;
2. Travel speed;
3. Available route options;
4. Path choice; and
5. Travel flow.

The consideration of delay-time needs to factor in the time from ignition to detection and from detection to alarm.

6.5.1 Delay Time Before Occupants Begin to Evacuate

Studies have shown that the time occupants delay before beginning to evacuate can vary by type of occupancy, whether people are awake or asleep, even time of year (weather-related issues). If people are asleep, there will be a period of time until they waken to an alarm or other cue. If they are intoxicated or have a hearing disability, there may be additional delays or they may not waken. It has also been shown that delay times often follow a log-normal distribution, so it is generally not appropriate for a given building/area population, to assume that a simple average delay time will suffice for the entire population. Quantitative data that can be used in modeling delay times, for various occupancies and occupant conditions, has been compiled in several tables in the fifth edition of the *SFPE Handbook of Fire Protection Engineering*, Chapter 64 [2].

6.5.1.1 Time from Ignition to Detection

The time from ignition to detection depends on the fire scenario and the means available for detection. For automatic heat or smoke detection systems, estimated time to detection can be calculated using fire and detector response calculation models. Alternatively, the first detection of a fire may be directly by a person or persons becoming aware fire cues. In practice both detection and warnings are processes involving a number of stages. A relatively simple situation is one in which automatic detection activates a general alarm to all occupants. More complex situations with longer detection times include larger buildings or transport systems with two staged alarm systems. In such cases the initial detection may be by an automatic detector or an individual person, leading to an alert to security. This is then typically followed by a sequence of investigation and reporting up and down a management chain, until a decision is made that a significantly serious fire exists to require a general alarm to be provided to affected occupants. For design purposes, detection time may be considered as the time to initial detection by an automatic system or a person.

6.5.1.2 Time from Detection to Alarm

For situations where automatic detection activates an immediate general alarm to occupants the alarm time is effectively zero. For situations where the fire is detected by a person, or where staged alarms systems are used, then the alarm time includes the subsequent period involving all pre-warning delays. Where first detection is by a person then the alarm time includes the delay until that person reports the fire and all subsequent delays in the reporting and response chain until a general alarm is provided to affected occupants. Where first detection is by an automatic detection system, the alarm time depends on the system protocol. For some systems the first detector activates a pre-warning to security staff, so that time to alarm then depends on the subsequent staff reporting and response chain. Some automatic systems have a fixed time-out delay before a general alarm is sounded, unless cancelled by staff. Other systems may provide a pre-alarm from first detection then a general alarm if a second detector is activated indicating a growing fire. Pre-warning delays may constitute a major part of total escape time and such delays have been an important contributory factor resulting in multiple deaths in a major fire incident in large buildings and transport systems [167, 168].

6.5.2 Travel Speed

Although some evacuation models calculate travel speeds or modify travel speeds based on local conditions such as crowdedness, or occupant characteristics such as age or disability, the engineer should be familiar with the ranges of speeds reported from drills and actual evacuations. Data is available for travel speed on stairs (up and down), on ramps, escalators, on horizontal surfaces, and with assistive devices (such as walkers or cane) and wheelchairs. This data is also compiled in Chapter 64 of the *SFPE Handbook of Fire Protection Engineering* [2].

6.5.3 Available Route Options

The pre-emergency availability of travel routes will be determined by the design of the building and the plans for the use of the building and its features by the occupants. The initial route availability will be an input requirement for any evacuation model or egress calculations. The availability of the routes during a fire or other emergency will be influenced by the fire scenarios, the expected development and spread of a fire, or could be altered by the engineer to test various design options.

6.5.4 Path Choice

Some evacuation models provide the options of shortest route or incorporate logic where route selection may be determined based on proximity, line of sight, waiting time or some other factor. The engineer will need to rely on engineering judgment in choosing options for exit route choice. In reality, the initial choice of exit path can be influenced by familiarity with the building, proximity to the exit, movement of others, conspicuousness of the exit, etc. During the course of an incident, local conditions can impact the occupants' ability to continue on their chosen path, and their likelihood of changing direction.

6.5.5 Travel Flow

The flow rate for the evacuation will depend on how crowdedness can occur on the travel paths. For example, in an occupancy such as an office with a trained population, the evacuation may begin after a very short delay, and the travel path will fill almost immediately, and the flow can be optimal (under the right conditions) or crowded. On the other hand, if the time to start evacuation is more staggered, or if the number of occupants in the building is relatively low, crowdedness may not occur and the occupants will move at an unimpeded but suboptimal flow rate. The engineer will need to choose an appropriate flow rate if doing a hand calculation or an evacuation model that requires a flow rate as an input variable. The flow rate must be consistent with the other characteristics of the occupants and scenario that would impact the flow conditions. Data on flows through doorways and on stairs is compiled in the fifth Edition of the *SFPE Handbook of Fire Protection Engineering* [2].

6.6 Sensitivity Analysis

A sensitivity analysis is a necessary step in the selection of design occupant scenarios, as it is in the selection of design fire scenarios. In a robust design, small changes in the specifics of the occupant characteristics (e.g., the proportion of the occupant population with disabilities, or the proportion choosing a certain exit option, etc.) should not result in significant changes in the outcome of the analysis. The ISO technical specification on the selection of design occupant scenarios recommends some specific variables for consideration in the sensitivity analysis – the number of occupants and their initial locations; combinations of occupant characteristics; initial responses of occupants; and initial exit route choices of occupants. It further specifies that the occupant characteristics of particular interest are: familiarity with the building (i.e., are they regular or transient users of the space); training level; age; abilities (i.e., sensory, cognitive and mobility); vulnerabilities; consciousness (i.e., awake, asleep, intoxicated or unconscious); role; and whether alone or with others.

Calculation of Effects of Fire Effluent

7

7.1 Toxicity Analysis Methods

7.1.1 Ct Product and Fractional Effective Dose

For the majority of toxic products in a fire atmosphere, a given toxic endpoint such as incapacitation or death occurs when the victim has inhaled a particular Ct product dose of toxicant. In order to make some estimate of the likely toxic hazard in a particular fire, it is, therefore, necessary to determine at what point the victim will have inhaled a toxic dose. A practical method for making this calculation is the concept of *fractional effective dose* (FED). The *Ct* product doses for small periods of time during the fire are divided by the *Ct* product dose causing the toxic effect, as shown in the following equation:

$$\mathbf{FED} = \frac{\textbf{Dose received at time t(Ct)}}{\substack{\textbf{Effective Ct dose to cause incapacitation} \\ \textbf{or death}}} \tag{7.1}$$

The fractional effective dose for each toxicant is, then, summed during the exposure until the fraction reaches unity. Unity is the point at which the toxic effect is predicted to occur (e.g. incapacitation, death).

For substances exhibiting a linear uptake (such as carbon monoxide at short exposure, high concentration), the denominator of the equation is a constant for any particular toxic effect. For substances deviating from linearity (such as carbon monoxide at long exposure (greater than 1 h), low concentration), the denominator for each time segment during the fire is the *Ct* product dose at which incapacitation would occur at the concentration during that time segment.

The concept of fractional effective dose can also be applied to sensory irritants and heat. For irritants the concept of *fractional irritant concentration* (FIC) has been developed, where

$$\mathbf{FIC} = \frac{\substack{\textbf{Concentration of irritant to which subject} \\ \textbf{is exposed at time (t)}}}{\substack{\textbf{Concentration of irritant required to} \\ \textbf{cause imparirment of escape efficiency}}} \tag{7.2}$$

For heat exposure, the fractional effective dose of heat acquired during exposure can be calculated by summing the radiant and convective fractions as

$$\mathbf{FED} = \int_{t_1}^{t_2} \left(\frac{1}{t_{Irad}} + \frac{1}{t_{Iconv}} \right) \Delta t \tag{7.3}$$

Calculation methods for specific toxicants and exposure conditions are discussed in more detail below.

Table 7.1 Ct product exposure doses for CO incapacitation by species at rest and light activity

	CO at rest (ppm-min)	CO light activity (ppm-min)
Human 70 kg	80,000–100,000	30,000-35,000
Baboon ~20 kg		34,000
Macaque 3–4 kg	38,000–40,000	27,000
Rat ~300 g	30,000–40,000	22,000–36,000

7.1.1.1 Carbon Monoxide

Incapacitation by CO depends on a dose accumulated over a period of time until a carboxyhemoglobin concentration is reached where compensatory mechanisms fail and collapse occurs. The simplest method for estimating time to incapacitation for an exposure to CO is the *Ct* product exposure dose method. This method is derived from experimental animal exposure data and then applied to calculation models on the assumption that incapacitation occurs at fixed *Ct* product exposure doses for any combination of concentration and exposure time. This method works reasonably well for short exposures to high CO concentrations under specific exposure conditions but makes no allowance for the effects of any physiological variables (specifically respiration). The *Ct* exposure doses for incapacitation in different species including humans at different levels of activity are shown in Table 7.1 [134].

Based on the data in Table 7.1, a *Ct* exposure dose of 35,000 ppm-min for incapacitation in average healthy humans engaged in light-moderate activity (fast waking) while escaping from a building is a reasonable value for fire hazard calculations. The advantage of this method is that it is extremely simple to use in fire hazard modeling, and it is considered to be valid for moderately active adult humans under the high CO concentrations and short exposure time scales usually occurring during fires. At this exposure dose it is predicted that approximately half an exposed population would be incapacitated. In order to protect more sensitive subjects such as the elderly or those with adverse health conditions including cardiovascular or lung conditions it is suggested that the exposure dose limit should be reduced by a factor of three [134, 139, 169, 170]. A caveat is that for applications outside these limitations (such as children, resting adults, long exposure periods at low CO concentrations or other situations involving differences in the variables listed) more complex modeling methods are necessary, such as the Stewart equation or Coburn-Forster-Kane (CFK) equation.

The Stewart equation is a simple analysis method which can be utilized to calculate the carboxyhemoglobin (COHb) concentration in the blood with user defined inputs of CO concentration (ppm CO), breathing rate (V_E (L/min)), and exposure time (t (min)).

$$\%\mathbf{COHb} = \left(\mathbf{3.317} * \mathbf{10^{-5}}\right)\left(\mathbf{ppm\ CO}\right)^{1.036}\left(\mathbf{V_E}\right)\left(\mathbf{t}\right) \tag{7.4}$$

This equation is valid for short exposures at high concentrations when the blood concentration is well below saturation level. Time to incapacitation is, then, predicted to occur between 30–40% COHb for healthy adult occupants (see Table 7.1).

When the inhaled concentration is high compared to that in the blood (as during short duration, high concentration exposures such as those that occur in flaming fires) the departure from linear uptake is not significant. However, over long periods at lower concentrations as equilibrium is approached, uptake deviates considerably from linearity. Additionally, the Stewart equation does not account for differences in body weight or lung capacity which is an important factor in children. In these cases, the Coburn-Forster-Kane (CFK) equation is most appropriate. It takes into account a range of variables, including RMV, body size, exposure duration, and parameters related to lung and blood physiology. More detailed information about the CFK equation can be found in the *SFPE Handbook of Fire Protection Engineering*. When these various factors are known, this equation enables accurate predictions of CO uptake to be made that agree well with experimental data.

The Stewart and CFK equations enable reasonably good predictions of time to incapacitation or death for short (less than 1 h) or long (greater than 1 h) exposures, respectively, to constant concentrations of CO in air. In fires, however, victims are exposed to concentrations of CO that change during the course of the fire. The basic rule to apply is that fluctuations are unlikely to cause the COHb concentration to deviate from that predicted by assuming the constant average concentration throughout, providing the CO concentration is on a rising trend, is stable, or is well above the equilibrium concentration with the blood COHb.

Although the average CO concentration during a fire exposure can be used to predict COHb concentration and time to incapacitation, another useful concept for predicting incapacitation is a variation on the *Ct* exposure dose method. Since the *Ct* "dose" actually represents the COHb concentration in the blood, the fractional dose would be better represented by the ratio of the COHb concentration at time, *t*, with the COHb concentration known to cause incapacitation rather than by simple *Ct* product ratios. Therefore, the Stewart equation can be rewritten in the form of COHb ratios, requiring only knowledge of the CO concentration and the exposure time as follows:

7.1 Toxicity Analysis Methods

Table 7.2 Values for V and D based on level of activity

Activity level of subject	V (L/min)	D (%COHb)
Resting or sleeping	8.5	40
Light work—Walking to escape	25	30
Heavy work—Slow running, walking up stairs	50	20

$$\mathbf{F_{I_{CO}}} = \mathbf{3.317 * 10^{-5}} \, \mathbf{[CO]^{1.036}} \, \mathbf{(V)(t)/D} \tag{7.5}$$

where [CO] is the carbon monoxide concentration (ppm v/v 20 °C), *V is the vo*lume of air breathed each minute (L), *t is the e*xposure time (min), and *D is the e*xposure dose (percent COHb) for incapacitation. The values in Table 7.2 may be taken for *V* and *D*.

Incapacitation is assumed to occur when the fractional dose reaches 1.0. The suggested default case for escaping occupants would be that of light work. This concept of *Ct* product fractional dose is also useful for predicting incapacitation from other fire products and combinations of products, as will be discussed later in this section.

The advantage of expressing CO exposure dose as a %COHb ratio rather than a ppm.min ratio is that the former calculates the actual CO dose inhaled, taking into account the main variables affecting uptake, while the ppm.min ratio only represents the dose available to be inhaled without allowing for these variables. Also, the calculated %COHb can be directly compared with the actual %COHb concentrations measured in the blood of fire survivors and decedents.

7.1.1.2 HCN

In practice, it is considered that time to incapacitation in the average adult human engaged in light physical activity would be similar to that in a resting monkey. On this basis, the following equation derived from primate data can be utilized to estimate time to incapacitation from hydrogen cyanide exposure [134, 139] for an adult human engaged in light activity (such as walking to escape from a fire):

$$t_{ICN} = \frac{1.2 * 10^6}{[CN]^{2.36}} \tag{7.6}$$

where t_{ICN} is the time to incapacitation (min) and [CN] is the inhaled HCN concentration (ppm). In a case where physical activity levels vary, the effects of different levels of respiratory ventilation can be accounted for using the following equation:

$$t_{ICN} = \frac{2.43 * 10^7}{[CN]^{2.36} \, V_E} \tag{7.7}$$

where V_E is the respiratory ventilation rate (L/min).

A method for estimating fractional dose to incapacitation from hydrogen cyanide was developed for the rat by Hartzell et al. [171]. In that model the *Ct* product over short periods of time is expressed as a fraction of the *Ct* product required to cause incapacitation at a specified concentration. The fractions for each short time interval are summed until the fraction reaches unity, which indicates incapacitation. This approach should enable reasonable predictions of time and dose to incapacitation provided that the HCN concentration is stable or increasing. This approach can be used to derive a fractional dose model for humans based on the time to incapacitation equation discussed above. The default expression recommended for fire engineering design applications uses the expression for resting primates as a predictor of active adult humans. Therefore, for a short exposure time, *t,* to a given HCN concentration

$$\mathbf{F_{ICN}} = \frac{\mathbf{(ppm \ HCN)(t)}}{\mathbf{(ppm \ HCN)(t_{I_{CN}})}} \tag{7.8}$$

where $\mathbf{F_{ICN}}$ is the fraction of an incapacitating dose. Assuming $t = 1$ min, the equation simplifies to

$$F_{ICN} = \frac{1}{t_{ICN}} \tag{7.9}$$

and provides the following fractional incapacitating dose expressions (basic expression for light activity and general expression with V_E as an input parameter):

$$F_{ICN} = \frac{[CN]^{2.36}}{1.2*10^6} \, t \tag{7.10}$$

$$F_{ICN} = \frac{[CN]^{2.36} \, V_E}{2.43*10^7} \, t \tag{7.11}$$

If the fractional doses per minute, F_{ICN}, are summed throughout the exposure, the dose and time to incapacitation can be predicted.

Example #1
A subject is exposed to 90 ppm HCN for 15 min, then to 180 ppm HCN for 3 min.

HCN concentration (ppm)	F_{ICN} for 1 min from Eq. (7.10)	$F_{ICN} \times$ time
90	0.03	0.51
180	0.18	0.53
$\sum F_{ICN}$ at 18 min		1.04

Therefore, incapacitation is predicted between 17 and 18 min.

7.1.1.3 Hypoxia
Using data from Luft, [172] it is possible to derive an equation that gives a reasonable prediction of time to loss of consciousness for a victim exposed to a hypoxic fire environment [134]. As with exposure to HCN, time to incapacitation for exposure to low-oxygen concentrations is not linear, since short exposures to severe hypoxia cause incapacitation very rapidly, and long exposures to modest hypoxia have little effect. In attempting to predict time or dose to incapacitation or death for a subject exposed to changing oxygen concentrations, it is therefore necessary to apply a weighting factor to allow for these deviations from ideal behavior. As with HCN this may be achieved by using the fractional effective dose concept as follows:

For a constant level of hypoxia, the time to incapacitation (t_{I_o}) due to oxygen depletion is given by

$$t_{I_o} = \exp[8.13 - 0.54 \, (20.9 - \%O_2)] \tag{7.12}$$

where $\%O_2$ is the oxygen exposure concentration. For short exposure time, t, to a given level of oxygen vitiation

$$F_{I_o} = \frac{(20.9 - \%O_2) \, (t)}{(20.9 - \%O_2)(t_{I_o})} \tag{7.13}$$

where F_{I_o} = fraction of an incapacitating dose of hypoxia. When $t = 1$ min, the equation simplifies to

$$F'_{I_o} = \frac{1}{t_{I_o}} \tag{7.14}$$

If the fractional doses per each minute are summed throughout the exposure, the dose and time to incapacitation can be predicted.

Example #2
A subject is exposed to a concentration of 10 percent oxygen for 5 min followed by 7.8% oxygen for 1.5 min.

7.1 Toxicity Analysis Methods

For 10% O_2

$$t_{I_0} = \exp[8.13 - 0.54(20.9 - 10)]$$

$$1/t_{I_0} = 0.11$$

For 7.8% O_2

$$t_{I_0} = \exp[8.13 - 0.54(20.9 - 7.8)]$$

$$\frac{1}{t_{I_0}} = 0.35$$

$$F_{I_0} = (0.11 \times 5) + (0.35 \times 1.5) = 1.08$$

Therefore, loss of consciousness is predicted at 6.5 min.

7.1.1.4 Carbon Dioxide

For asphyxiant gases such as CO or HCN, it is likely that the increased uptake resulting from carbon dioxide induced hyperventilation will significantly reduce time to incapacitation and death. An average curve developed from three data sources allows for the development of an equation to represent the enhanced uptake of other asphyxiant gases as follows:

$$\mathbf{VCO_2} = \frac{\exp\left(\mathbf{0.1903\%CO_2 + 2.0004}\right)}{\mathbf{7.1}} \tag{7.15}$$

Using a respiratory rate of 6.8 L/min (suggested figure for a resting respiratory rate at the background CO_2 concentration), and allowing for certain inefficiencies in CO uptake rate [134] the equation can be simplified to

$$\mathbf{VCO_2} = \exp\left(\frac{[\mathbf{CO_2}]}{\mathbf{5}}\right) \tag{7.16}$$

As with HCN and low-oxygen hypoxia, intoxication by carbon dioxide does not follow a linear trend (Ct for 10% CO_2 = 20 percent-min, Ct for 5% CO_2 = 175 percent-min). Based on research data, an expression predicting time to incapacitation $t_{I_{co_2}}$ can be derived as follows:

$$t_{I_{co_2}} = \exp(\mathbf{6.1623 - 0.5189\%CO_2}) \tag{7.17}$$

Using the fractional-dose concept previously described for HCN and hypoxia, it is possible to predict a dose to incapacitation provided that the CO_2 concentration is stable or increasing, as follows:

$$\mathbf{F_{I_{co_2}}} = \frac{(\mathbf{\%CO_2})(\mathbf{t})}{(\mathbf{\%CO_2})\left(\mathbf{t_{I_{co_2}}}\right)} \tag{7.18}$$

where $\mathbf{F_{I_{co_2}}}$ is the fraction of an incapacitating dose. When $t = 1$ min, the equation simplifies to

$$\mathbf{F_{I_{co_2}}} = \frac{\mathbf{1}}{\mathbf{t_{I_{co_2}}}} \tag{7.19}$$

If the fractional doses per minute are summed throughout the exposure, the dose and time to incapacitation can be predicted.

Example #3

A subject is exposed to a concentration of 5% CO_2 for 20 min, followed by 9% CO_2 for 2 min.

$$\text{For } 5\% CO_2, t_{I_{CO_2}} = 35.44; 1/t_{I_{CO_2}} = 0.03$$

$$\text{For } 9\% CO_2, t_{I_{CO_2}} = 4.45; 1/t_{I_{CO_2}} = 0.22$$

$$F_{I_{CO_2}} = 0.0282 \times 20 + 0.2247 \times 2 = 1.04$$

Severe incapacitation with probable loss of consciousness is, therefore, predicted at approximately 22 min.

7.1.1.5 Irritants/Smoke

In order to assess the visual obscuration effects of smoke, a concept of fractional effective concentration (FEC) has been developed, whereby the smoke concentration is expressed as a fraction of the concentration considered to significantly affect escape efficiency. The default limit levels suggested in Purser and McAllister [134] and ISO TR16738 [173] are optical density per meter (OD/m) of 0.2 for small enclosures and 0.08 for large enclosures. These values were selected on the basis that occupants of built enclosures should be able to see an exit in order to escape effectively. The values are not intended to reflect any true endpoint condition so that design limits may be selected depending the specific application or risk level deemed acceptable. In addition, it is considered that for smoke densities up to 0.8 OD/m the concentrations of other toxic gases in the smoke will be relatively harmless for periods of up to an hour, while for densities up to 0.2 OD/m they will be insufficient incapacitation within approximately 20 min. Refer to Sect. 7.2 for a full discussion on visibility.

$$\mathbf{FEC_{smoke}} = \left(\frac{\mathbf{OD}}{\mathbf{m}}\right)/\mathbf{0.2} \text{ for small enclosures (5m) or } \left(\frac{\mathbf{OD}}{\mathbf{m}}\right)/\mathbf{0.08} \text{ for large enclosures (10m)} \tag{7.20}$$

If the total FEC_{smoke} reaches unity, then it is predicted that the level of visual obscuration would be sufficient to seriously affect escape attempts. In order to derive calculation expressions for the relationship between smoke density and walking speed for non-irritant or irritant smoke, it is possible to perform a variety of fits to data sets from Jin [158, 174] and Frantzich and Nilsson [175] (see Purser and McAllister) [134]. For a simple deterministic calculation an expression fitted to Jin's irritant data and the Frantzich and Nilsson [175] data is suggested as follows:

$$\mathbf{W_{smoke} \ (m/s) = -0.1364 \ Ln \ \alpha_k + 0.6423} \tag{7.21}$$

where W_{smoke} is the walking speed in moderately irritant smoke and α_k is the extinction coefficient (see Table 7.2). The standard deviation around the best fit curve is ±0.157 m/s. This expression is suggested for smoke typical of fires in built enclosures involving generic mixed fuels, which is likely to be moderately irritant.

Additional expressions utilizing optical density rather than extinction coefficient for nonirritant (W_{nonirr}) and irritant smoke (W_{irr}) are

$$\mathbf{W_{nonirr} \ (m/s) = 5.5 \ (smoke \ OD/m)^3 - 5.6 \ (smoke \ OD/m)^2 - 0.2 \ (smoke \ OD/m) + 1.2} \tag{7.22}$$

$$\mathbf{W_{irr} \ (m/s) = -27.3 \ (smoke \ OD/m)^2 + 2.2 \ (smoke \ OD/m) + 1.2} \tag{7.23}$$

For nonirritant and moderately irritant smoke, the minimum speed in darkness is 0.3 m/s in cases where occupants continue to move (are willing to continue moving through smoke). For highly irritant smoke, it is considered more likely that movement may cease.

Where significant concentrations of irritant gases are considered likely to be present (for example in electrical fires or those involving transport vehicles), the combined effects of sensory irritants should also be considered in a hazard evaluation. It is difficult to quantify irritant effects exactly, because the database on the effects of individual irritants or irritant mixtures on escape behavior in humans is poor. Since the effects rely on a continuum of severity, there are no precise endpoints. In order to assess the combined effects of irritants, a concept of fractional irritant concentration (FIC) has been developed, whereby the concentration of each irritant present is expressed as a fraction of the concentration considered to be severely irritating. The

7.1 Toxicity Analysis Methods

Table 7.3 Irritant concentrations of common fire gases

Gas	Concentration predicted to impair escape in half the population (ppm)	Concentration predicted to cause incapacitation in half the population (ppm)
HCl	200	900
HBr	200	900
HF	200	900
SO_2	24	120
NO_2	70	350
CH_2CHO (acrolein)	4	20
HCHO (formaldehyde)	6	30

FICs for each irritant are then summed to give a total FIC. If the total FIC reaches unity, then it is predicted that the smoke atmosphere would be highly irritant, sufficient to slow escape attempts. If the total exceeds unity by a factor of approximately four or more, then it is likely that escape would be prevented and possible that collapse might occur due to static hypoxia from bronchoconstriction or laryngeal spasm. On the basis of available data, current estimates of the concentrations of each gas likely to be highly irritant are as shown in Table 7.3.

It is predicted that each gas at the above concentrations is likely to be sufficiently irritant to affect escape efficiency in the majority of subjects and may cause incapacitation in susceptible individuals. A factor of 0.3 FEC for escape impairment should allow for safe escape of nearly all exposed individuals.

On the basis of the assumption that all irritants capable of damaging lung tissue are additive in their effects, the overall irritant concentration would be represented as

$$\mathbf{FIC} = \mathbf{FIC_{HCl}} + \mathbf{FIC_{HBr}} + \mathbf{FIC_{HF}} + \mathbf{FIC_{SO_2}} + \mathbf{FIC_{NO_2}} + \mathbf{FIC_{CH_2CHO}} + \mathbf{FIC_{CH_2O}} + \sum \mathbf{FIC_x} \qquad (7.24)$$

where $\sum FIC_x$ = FICs for any other irritants present.

A considerable degree of variability is likely in individual susceptibility within the population and there is a degree of uncertainty as to which tenability thresholds should be recommended. In the recently published international standard for the estimation of time available for escape (ISO 13571) [169] the chosen tenability endpoint has been that of predicted incapacitation (rather than escape impairment), was defined as sublethal effects that would render persons of average susceptibility incapable of effecting their own escape. The endpoint and predicted endpoint concentrations are very similar to those in the right-hand column of Table 7.6 with a suggestion that design tenability limits might be set at 0.33 times these concentrations to allow for more sensitive subpopulations.

7.1.1.6 Overall FED Expression for Asphyxia

In general, although there is evidence for interactions between toxic fire gases, whether these are likely to be important in practice depends on the composition of actual fire atmospheres. For most practical situations, it is considered that HCN is unlikely to be the main driver of time to incapacitation providing the nitrogen content of the burning fuel does not exceed 1% by mass. The composition of fire atmospheres is such that CO is frequently the most important toxic product, and the most important interaction is then increased rate of CO uptake due to hyperventilation caused by CO_2. The additional effects of HCN hypoxia contribute to the effects of CO-induced asphyxia and may significantly reduce time to incapacitation when fuels contain more than 1% nitrogen, and the HCN concentration exceeds approximately 50 ppm. Low oxygen hypoxia is likely to be a minor term for situations where subjects are exposed to growing compartment fires. However, it may constitute a major term if subjects are suddenly exposed to a smoke atmosphere containing less than approximately 12% oxygen (e.g. when opening a door to a compartment on fire). Similarly, sudden exposure to a CO_2 concentration exceeding 7% could itself result in rapid intoxication and collapse.

On this basis, the simplified fractional dose equation for asphyxiation for an adult engaged in light work would be

$$\mathbf{F_{IN}} = [(\mathbf{F_{ICO}} + \mathbf{F_{ICN}} + \mathbf{FIC})\ \mathbf{VCO_2} + \mathbf{F_{I_o}}]\ \text{or}\ \mathbf{F_{ICO2}} \tag{7.25}$$

where

$\mathbf{F_{IN}}$ = Fraction of an incapacitating dose of all asphyxiant gases
$\mathbf{F_{ICO}}$ = Fraction of an incapacitating dose of CO
$\mathbf{F_{ICN}}$ = Fraction of an incapacitating dose of HCN (and nitriles, corrected for NO_2)
$\mathbf{FIC}$ = Fraction of an irritant dose contributing to hypoxia
$\mathbf{VCO_2}$ = Multiplication factor for CO_2-induced hyper-ventilation
$\mathbf{F_{I_o}}$ = Fraction of an incapacitating dose of low-oxygen hypoxia
$\mathbf{F_{ICO2}}$ = Fraction of an incapacitating dose of CO_2

Example #4

Applying the expressions for the fractional incapacitating dose of each gas to the data in Table 7.4, the total fractional dose of all asphyxiant gases for each minute during the fire has been calculated in Table 7.5.

Table 7.4 Average concentrations of asphyxiant gases each minute during the first 6 min of the single armchair room burn

Time (min)	1	2	3	4	5	6
CO ppm	0	0	500	2000	3500	6000
HCN ppm	0	0	0	75	125	174
$CO_2\%$	0	0	1.5	3.5	6	8
$O_2\%$	20.9	20.9	19	17.5	15	12

Table 7.5 Average concentrations of asphyxiant gases each minute during the first 6 min of the single armchair room burn

Time (min)	1	2	3	4	5
$\mathbf{F_{ICO}}$	0	0	0.017	0.074	0.130
+ $\mathbf{F_{ICN}}$	0	0	0.000	0.002	0.074
$\times\ \mathbf{V_{CO_2}}$	0	0	1.442	2.376	4.434
=	0	0	0.025	0.228	0.905
$\mathbf{F_{IO}}$	0	0	0.001	0.002	0.007
= Total	0	0	0.026	0.230	0.912
Running total ($\mathbf{F_{IN}}$) or	0	0	0.026	0.256	1.168
$\mathbf{F_{ICO2}}$	0	0	0.005	0.013	0.047
Running total ($\mathbf{F_{IN}}$)	0	0	0.005	0.018	0.065

Incapacitation (loss of consciousness) is predicted at 5 min when the fractional incapacitating dose exceeds unity ($F_{IN} = 1.2$).

7.1.1.7 Heat

During a fire incident subjects may be exposed mainly to radiant heat (for example when walking below a hot smoke layer or past a flame), mainly to convected heat (for example in an overheated room) or to a combination of radiant and convected heat (for example when immersed in a hot smoke plume). Babrauskas [176] suggests a tenability limit of 2.5 kW/m². At a level of 2.5 kW/m², Buettner [177] and Simms and Hinkley [178] found that pain occurred within approximately 30–60 s of exposure. Tolerance times increase exponentially as the radiant flux decreases; Mudan and Croce [179] reported a critical heat flux of 1.7 kW/m², below which no pain was experienced, regardless of the exposure duration. Radiant heat above 2.5 kW/m² will cause skin pain, followed by burns within a few seconds. For situations where occupants are required to pass under a hot smoke layer in order to escape, a radiant flux of 2.5 kW/m² corresponds to a hot layer temperature of approximately 200 °C. Above this threshold, time (minutes) to different endpoints for effects of exposure to radiant heat (t_{Irad}), at a given radiant flux of q (kW/m²) is given as

7.1 Toxicity Analysis Methods

Table 7.6 Limiting conditions for tenability caused by convective heat

Intensity	Tolerance time
<60 °C at 100% saturation	>30 min
100 °C at <10% water content by volume	12 min
120 °C at <10% water content by volume	7 min
140 °C at <10% water content by volume	4 min
160 °C at <10% water content by volume	2 min
180 °C at <10% water content by volume	1 min

$$t_{I_{rad}} = \frac{r}{q^{1.33}} \tag{7.26}$$

where r is the radiant heat exposure dose [$(kW \cdot m^{-2})^{4/3}$ min] required for any given endpoint. It is proposed that a figure of 1.33 $(kW \cdot m^{-2})^{4/3}$ is used to represent a pain tolerance threshold and 10 $(kW \cdot m^{-2})^{1.33}$ is used to represent a threshold for incapacitation and serious injury (potentially fatal for those aged over 65 years).

Research suggests a tolerance limit of approximately 120 °C for unprotected skin exposed to convective heat. The tolerance time is dependent upon the humidity of the air. Table 7.6 provides a summary of tolerance times for various temperatures and humidity ranges.

For exposures, of up to 2 h, to convected heat from air containing less than 10% by volume of water vapor, the time (minutes) to incapacitation $t_{I_{conv}}$ at a temperature T (°C) is calculated as:

$$t_{I_{conv}} = 5 * 10^7 \, T^{-3.4} \tag{7.27}$$

This expression applies best when humidity approaches 100%. It is, therefore, a somewhat under-conservative expression for exposure to higher temperatures and somewhat over-conservative expression for exposures on the low-temperature end. The following expressions have been developed for the mid-humidity case, giving a better fit to the empirical data. The tolerance time t_{tol} (minutes) under mid-humidity conditions is then given by

$$t_{tol} = 2 * 10^{31} \, T^{-16.963} + 4 * 10^8 \, T^{-3.7561} \tag{7.28}$$

where T is the ambient temperature (°C). This expression is considered suitable for calculating tolerance time as a possible tenability limit for design purposes. For other applications, such as probabilistic risk assessments other endpoints may be required. Endpoints may include when serious injury from severe hyperthermia or second-degree burns (considered to represent a point of incapacitation for the average occupant) may occur. Based on the hyperthermia data and the relationship between heat doses causing pain and those causing serious injury, similar expressions have been derived for predicting time to serious injury. Time (minutes) to serious injury or severe incapacitation is represented as:

$$t_{injury} = 5 * 10^{22} \, T^{-11.783} + 3 * 10^7 \, T^{-2.9636} \tag{7.29}$$

These expressions are related to exposure to heated air with less than 10% water content by volume.

As with toxic gases, the body of a fire victim may be regarded as acquiring a "dose" of heat over a period of time during exposure, with short exposure to a high radiant flux or temperature being more incapacitating than a longer exposure to a lower temperature or flux. The same fractional incapacitating dose model as with the toxic gases may be applied and, providing that the temperature in the fire is stable or increasing, the fractional dose of heat acquired during exposure can be calculated by summing the radiant and convective fractions as shown:

$$FED = \int_{t_1}^{t_2} \left(\frac{1}{t_{I_{rad}}} + \frac{1}{t_{I_{conv}}} \right) \Delta t \tag{7.30}$$

7.1.2 Life Threat Hazard Analysis

In the previous section, the various elements of a physiological FED model for predicting time to incapacitation of occupants during full-scale fires have been summarized. The following section consists of a worked example including tenability calculations for all toxic and physical hazards: time-to-escape efficiency impairment from the effects of optical obscuration by smoke, time-to-escape efficiency impairment from sensory irritation, time to incapacitation by asphyxiant gases, time to incapacitation due to skin pain and burns from radiant and convected heat, and time to inhale a lethal exposure dose of lung irritants. The worked example utilizes data from an open room burn involving a rapidly growing fire in an armchair constructed from polystyrene with polyurethane covers and cushions.

Table 7.7 shows the input data for the fire and results for the hazard analysis. The endpoints of escape impairment or loss of tolerability (for smoke obscuration and irritants) and incapacitation (for heat and asphyxiant gases) are reached when the line for each parameter crosses 1.0. Higher FECs and FEDs indicate more severe effects. For irritancy, incapacitation is predicted at FEC_{irr} values of approximately 5–10.

Table 7.7 Inputs for hazard analysis

Gas concentrations						
Each minute	1	2	3	4	5	6
Smoke (OD/m)	0.1	0.2	0.5	1.5	3.0	3.5
HCl (ppm)	10	50	150	200	250	200
Acrolein (ppm)	0.4	0.8	2.0	6.0	12.0	14.0
Formaldehyde (ppm)	0.6	1.2	3.0	9.0	18.0	21.0
CO (ppm)	0	0	500	2000	3500	6000
HCN (ppm)	0	0	50	150	250	300
CO_2 (%)	0	0	1.5	3.5	6.0	8.0
O_2 (%)	20.9	20.9	19.0	17.5	15.0	12.0
Temp (°C)	20	65	125	220	405	405
Heat flux (kW/cm^2)	0	0.1	0.4	1.0	2.0	2.5
FEC_{smoke}	0.50	1.00	2.50	7.50	15.00	17.50
FIC_{HCl}	0.05	0.25	0.75	1.00	1.25	1.00
$FIC_{acrolein}$	0.10	0.20	0.50	1.50	3.00	3.50
FIC_{form}	0.10	0.20	0.50	1.50	3.00	3.50
Σ FIC	0.25	0.65	1.75	4.00	7.25	8.00
FED_{ICO}	0.00	0.00	0.02	0.07	0.13	0.23
FED_{ICN}	0.00	0.00	0.01	0.11	0.38	0.58
FLD_{irr}	0.00	0.00	0.00	0.00	0.01	0.04
VCO_2	1.00	1.00	1.35	2.01	3.32	4.95
FED_{IO2}	0.00	0.00	0.00	0.00	0.01	0.04
FED_{IN}	0.00	0.00	0.04	0.38	1.72	4.08
Σ FED_{IN}	0.00	0.00	0.04	0.42	2.14	6.23
FED_{rad}	0.00	0.00	0.00	0.00	2.54	2.54
FED_{conv}	0.00	0.02	0.19	1.57	15,55	15.55
FED_{heat}	0.00	0.02	0.19	1.57	18.10	18.10
Σ FED_{heat}	0.00	0.02	0.21	1.78	19.88	37.98

The analysis is designed to predict the severity of each hazard and the time during the fire at which it becomes significant. The toxic gas concentrations, smoke optical density, temperature, and radiant heat flux have been averaged over each of the first 6 min of a theoretical furniture fire but are generally similar to conditions obtained in the smoke layer at head height in some experiments performed in ISO room tests. The analysis shows that the smoke obscuration is the first hazard confronting a room occupant. The level of obscuration exceeds the tenability limit for irritant smoke in a small enclosure after the second minute, with an FEC of 1. The second hazard to confront the occupant is irritancy. This becomes significant during the third minute, reaching an FIC of 1 just after 2 min. The tenability limit designed to protect vulnerable individuals (FIC 0.3) is exceeded approximately 1 min earlier. It is therefore predicted that after the second minute the level of obscuration and the irritancy of the smoke would be sufficient to impair and possibly prevent escape from the room due to difficulty in seeing and increasing pain in the eyes and respiratory tract. The effects of radiant and convective heat then become significant, crossing the tenability limit during the fourth minute and reaching an FED_{heat} value of 1.78, so that it is predicted that a room occupant

7.1 Toxicity Analysis Methods

would suffer severe skin pain and then some burns due to the effects of convective heat, then producing potentially lethal full thickness burns during the fourth or fifth minute.

During the fifth minute the radiant flux reaches the tenability limit of 2.5 kW/m^2, so that skin pain would be predicted within seconds due to radiation alone, were it not that the temperature has already exceeded the limiting exposure dose. Also, during the fifth minute the FED$_{IN}$ reaches 2.14, predicting that anyone breathing the smoke would lose consciousness due to asphyxia and might die after 6 min. The level for exposure to asphyxiants considered to provide protection for vulnerable sub-populations (FED$_{IN}$ of 0.1) is crossed at approximately 3.5 min, half a minute before the FED$_{IN}$ reaches 1. The exposure dose of irritants is very small during the first 6 min of the fire, so that there should be little danger of post-exposure lung damage. An important point about all these parameters is that the FIC and FED curves are rising very steeply after the tenability threshold (FIC or FED of 1) is crossed. This means that even if the true exposure concentrations or exposure doses required to cause incapacitation were higher than the tenability limits chosen there would be little effect on predicted time to incapacitation.

The overall prediction is that for this fire, escape would become difficult during the third minute and incapacitation could occur due to the effects of irritant smoke. A person remaining in the room after this time would suffer severe pain and burns after 4 min, which would probably be lethal. In this analysis it is assumed that the head of a room occupant would be in the smoke at all times. In practice, if the room doorway was open, the hot, effluent-rich layer would descend from the ceiling to a level probably between 1 and 1.2 m above the floor as the chair reached its peak burning rate. A more sophisticated analysis could allow for the possibility that a room occupant might be at, or move to, a lower level in the room. If the height of the smoke layer with time is measured, then it is possible to allow for this in the calculation.

7.1.3 Typical Production Levels Based on Fire Type

Based on product composition and toxic hazard, it is possible to distinguish four basic types of fires for use in hazard analyses.

1. Non-flaming thermal decomposition/smoldering fires
2. Early/developing flaming fires
3. Small oxygen vitiated flaming fires (pre-flashover, under-ventilated compartment fires)
4. Fully developed fires (post-flashover fires)

Each fire type produces different concentration of toxic fire gases, and therefore, time to incapacitation can vary based on the exposure conditions. Table 7.8 provides a summary of the fire types and characteristics production ranges that could be utilized for life safety evaluations.

Table 7.8 Classification of toxic hazards in fires as revealed by large-scale fire simulation tests

Fire	Rate of growth	CO$_2$/CO	Toxic hazard	Time to incapacitation	Escape time available
1. Smoldering/non-flaming: Victim in room of origin or remote	Slow	~1	CO 0–1500 ppm low O$_2$ 15–21% irritants, smoke	Hours	Ample if alerted
2. Well ventilated flaming: Victim in room of origin	Rapid	1000 decreasing toward 50	CO 0–0.2% CO$_2$ 0–10% low O$_2$ 10–21% irritants, heat, smoke	A few minutes	A few minutes
3. Small vitiated flaming: Victim in room of origin or remote	Rapid, then slow	<10	CO 0.2–4% CO$_2$ 1–10% O$_2$ <12% HCN to 1000 ppm, irritants, heat, smoke	A few minutes	A few minutes
4. Fully developed: (post-flashover) victim remote	Rapid	<10	O$_2$ 0–3% in upper layer flowing from fire CO 0–3%[a] HCN 0–1000 ppm some irritants, smoke, and possibly heat	<1 min near fire, elsewhere depends on degree of smoke dilution	Escape may be impossible or time very restricted. More time at remote locations

[a]Concentrations depend on position relative to fire compartment

Design engineers should take into consideration the possible toxicant yields and production rates for various types of fire when performing a hazard analysis.

7.1.4 Susceptible Populations

The physiological algorithms are based partly on experimental data and reported effects in humans and partly on animal studies. These methods involve either the exposure dose or concentration predicted to produce a given effect on humans exposed to fire effluent. However, the effects are based on data for healthy young, adult animals or humans. The overall human population contains a number of subpopulations, however, which exhibit greater sensitivity to various fire effluent toxicants. Typically, these sensitivities are due to compromised cardiovascular and pulmonary systems.

Two of the largest such subpopulations are the elderly and the approximately 15% of children and 5% of adults who are asthmatic. The elderly, and particularly those with impaired cardiac perfusion, are particularly susceptible to asphyxiant gases. Thus, the average lethal carboxyhemoglobin (COHb) concentration in adults dying in fires or from accidental CO exposure is lower in the elderly. Asthmatics and sufferers of other lung conditions, such as chronic bronchitis and reactive airways dysfunction syndrome, are particularly susceptible to bronchoconstriction on even brief exposure to very low concentrations of irritants; with distress, severely reduced aerobic work capacity, collapse, and resulting death (depending on the sensitivity of the individual and the severity of the exposure).

It is the objective of fire safety engineering to ensure that essentially all occupants, including sensitive subpopulations, should be able to escape safely without experiencing or developing serious health effects. Thus, safe levels for exposure of the human population to fire effluent toxicants must be significantly lower than those determined from experiments with uniformly healthy animal or even human surrogates.

The toxicity endpoints from the functions presented in this section are all predicted to represent the median of the distribution of exposure doses or concentrations resulting in a given toxicity endpoint. Since individual susceptibility varies in the population, it is considered that approximately 11.3% of the population is likely to be susceptible below an FED of 0.3 (see ISO 13571). Approximately 90% of the population is considered to be susceptible below an FED of 1.3. For this reason, it will be necessary for the user to select an FED value to protect an acceptable proportion of vulnerable subpopulations (for example, an FED of 0.3 or some other value). It should be noted, however, that due to the rapid (t^2) rate of increase of smoke and asphyxiant gas concentrations in most flaming fires, variations in individual susceptibility and uncertainties in prediction of incapacitating doses tend to have relatively minor effects on predicted times to incapacitation.

7.2 Background and Guidance on Reduced Visibility Conditions

The basis for our current knowledge on human subject visibility in smoke emanated from the work of Jin and Yamada in the 1970's. Since Jin's work is widely referenced it is important to understand the background of his experiments and limitations of his work. Jin conducted and coauthored several studies that discuss the various effects of smoke on visibility. One of his key studies involved the determination of the visibility of signs as determined by human subjects. In this study both light emitting signs and light reflecting signs were observed in a smoke-filled chamber from outside through a glass window. As a surrogate for the signs Jin used light emitting circular discs and externally illuminated reflective discs. Both white smoke from a smoldering source and black smoke from a flaming source were used to evaluate difference in smoke composition. It is important to recognize that this early work by Jin did not expose the subjects to the smoke and the experimental visual environment is not comparable to the diverse field of view a human subject would have in a travel path.

The relationship between visibility and smoke density resulting from this work by Jin is given as:

$$V = \frac{\mathbf{Const.}}{\mathbf{C_s}} \tag{7.31}$$

Where V is the visibility of signs at the obscuration threshold (m), $\mathbf{C_s}$ is the smoke density expressed by the extinction coefficient (m^{-1}), and **Const.** is a constant that changes relative to whether the sign is light-emitting or reflective. The visibility expressed above is the distance where the sign just begins to become recognizable as an object. The extinction coefficient can be converted into optical density (OD/m) by dividing it by 2.303 (e.g. an extinction coefficient of 2.0 is equivalent to an optical density of 0.87).

7.2 Background and Guidance on Reduced Visibility Conditions

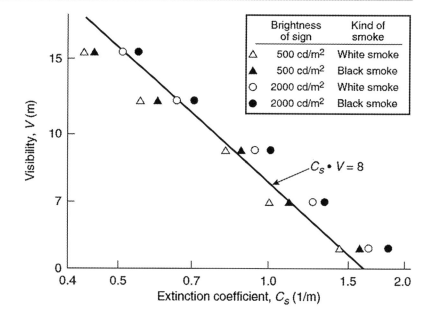

Fig. 7.1 Visibility vs. extinction coefficient (log-log of Jin's visibility relationships) [181]

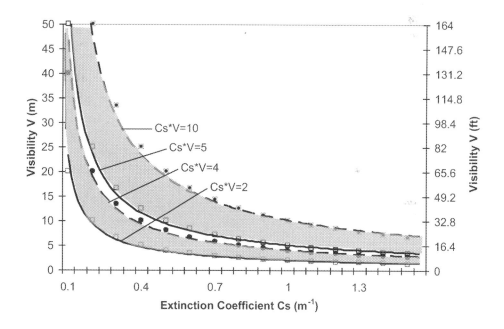

Fig. 7.2 Non-log plot of Jin's visibility relationship

Jin et al. found that for reflecting signs, the empirically determined constant was in a range of 2–4; the range is a function of the illumination of the sign and color of the smoke. In addition to the visibility of reflective signs, Jin et al. [180] also suggested that this value is applicable for the visibility of other objects such as walls, floors, doors, and stairways. Physical features of a space, such as the walls, doors, and floors, are important markers that aid exiting. For light emitting signs, the empirically determined constant was in a range of 5–10. Jin [181] presents this data on a log-log graph of Visibility versus Extinction Coefficient as shown in Fig. 7.1 for light emitting signs.

Jin et al. originally plotted this data on a log-log graph of Visibility versus Extinction Coefficient. For ease of understanding, Jin's data can be expressed in a non-log manner as provided in Fig. 7.2. The graph demonstrates that at higher values of extinction coefficient result in lower values of visibility at the obscuration threshold. As the constant value is lowered, factors such as light-emitting versus reflective signage are taken into account. The irritancy of the smoke also needs to be considered in the engineering analysis (See Fig. 7.3).

Figures 7.1 and 7.2 did not address directly exposing subjects to smoke and gases from the smoldering/flaming sources used. Jin, however, performed another series of tests to determine the effects of direct exposure to smoky atmospheres. Using a 20 m long corridor filled with smoke, subjects were instructed to walk into one end and record the place in the corridor where

Fig. 7.3 Log-log plot of Jin's data: effects of irritant and non-irritant smoke

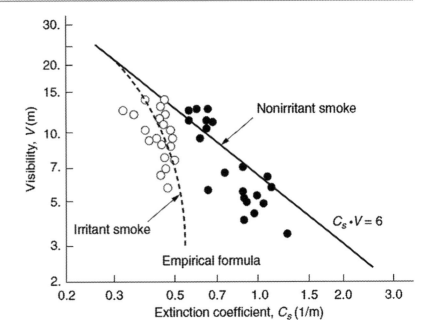

Fig. 7.4 Effects of irritant smoke and nonirritant smoke on walking speed

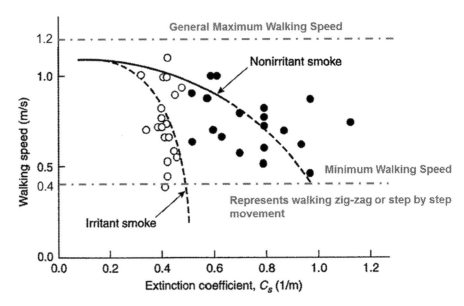

they could see a lighted exit sign or read the words at the opposite end of the corridor. The smoke in the corridor was either a highly irritant white smoke (wood cribs) or a less irritant black smoke (kerosene). Figure 7.3 below shows Jin's data and the effects of irritant and non-irritant smoke on human subjects. This data is based on asking the subjects to read the words on a lighted exit sign, meaning a light emitting source was used. Given that the data applies to a light emitting source, a correction can be made by adjusting the data by a factor of 2.5 to result in values for a light reflecting source. Jin's visibility data points from Fig. 7.3 for a light emitting source using an extinction coefficient of approximately 0.5 m^{-1} show visibility in the range of 5–7 m (16.4–22.9 ft). Correcting for reduced visibility with light reflecting sources the visibility would be approximately 2.5–3.5 m (8.2–11.5 ft). The adjustment factor of 2.5 is resultant from the ratio of constant values as established by Jin for light emitting versus light reflecting signs.

The data in Fig. 7.3 [181] clearly shows a visibility reduction due to irritant smoke not indicated with nonirritant smoke. Jin noted in the thick irritant smoke, subjects could only keep their eyes open for a short time and that tears were so frequent that they could not see the words on the sign. Jin opined that in this case when the signs are so simple or sufficiently familiar to the occupants to be recognized at a glance, this irritant effect of smoke may not cause so much trouble in locating the exits.

The 20-foot long corridor experiment also demonstrated the relationship between direct smoke exposure of irritant and nonirritant smoke. Figure 7.4 shows that smoke density and irritation both impact walking speed. The walking speed in the

7.2 Background and Guidance on Reduced Visibility Conditions

nonirritant smoke is shown to decrease gradually with increasing smoke density. For irritant smoke the effect is more pronounced than nonirritant smoke as the walking speed decreases rapidly after reaching an extinction coefficient of approximately 0.35/m; this rapid decrease in speed is similar to that noted in Fig. 7.3 for a decrease in visibility with irritant smoke. Jin noted that the sharp drop in walking speed is explained by the subject's movement; as they could not keep their eyes open, they walked inevitably in a zigzag motion or step by step along the side wall. In these studies, the concentration of irritants within the smoke is unknown; therefore, the application of this data to a large variety of fire scenarios should be considered in the engineering design.

Simply selecting any one individual's suggested visibility value or associated extinction coefficient does not afford an understanding of the relevance of the value selected. For example, Jin suggested the value of 0.5 m^{-1} extinction coefficient (converts to 0.22 OD/m) as a reasonable visibility limit for evacuees exiting from a public building. This equates to a conservative visibility distance for light reflecting situations of 4 m (13 ft). Jin's suggestion [313] stems from an experimental study using 49 subjects in a test chamber of 5 m by 4 m, with no windows and floor illumination of 30 lx. Half of the test subjects were Institute researchers who were introduced to the test chamber and briefed once on what to expect from the testing. Using white smoke from wood chips heated in a furnace, each subject was asked to thrust a metal stylus into holes of a device called a steadiness tester. Emotional variations were noted based on whether or not the stylus touched the hole edges and served as an indication of the effects of increasing smoke concentration in the room. From this experiment, Jin concluded that a value of 4 m (13 ft) visibility would be an appropriate limit that allows for safe escape for occupants familiar with a public building. The inference is that in a fire scenario occurring in a public building, when visibility declines to 4 m (13 ft), occupants will cease to move through a long corridor or abandon efforts to escape even without being overcome by heat or toxic gas exposure.

More recent studies on movement through smoke have been carried out in simple corridor settings (similar to the ones used by Jin et al.), in buildings and in road and rail tunnels [182–197]. A comparison between some of these experimental findings with Jin's work is presented in Purser and McAllister [134]. There is, however, a large variation in results yielding from different studies [198]; in addition, differences in experimental methodology, execution and documentation sometimes make the results difficult to directly combine in order to provide a more general description of peoples' movement during evacuation in smoke-filled environments. Thus, the little amount of available data, in combination with the difficulties to combine this data, may lead to an unwanted propagation of uncertainties in life safety analyses including evacuation in smoke. This has also been acknowledged in the past, both by developers of evacuation simulation software and fire safety designers. The consequence has been that the related uncertainty has been treated with crude and conservative assumptions regarding peoples' walking speed in smoke for different visibility levels.

Thus, in addition to the little amount of available data on peoples' walking speed in smoke, there has been a lack of knowledge and practice on how to use, represent and describe this data in practical RSET analyses. In 2015, a Swedish research project was initiated by the Swedish Transport Administration with the goal to summarize the current knowledge, and to describe and recommend how it can be used during practical application. Among other things, the project rendered a recommendation on how to represent peoples' walking speed in smoke in life safety verifications, depending on how the treatment of uncertainties is done in the analysis. The recommendation is based on the available (at that time) literature and data on human behaviour and movement in smoke filled environments, and builds on the information, data and conclusions that has been presented in a range of studies, of which the results have been deemed possible to combine in order to yield a more generalizable representation. A full account of the recommendation and the background is presented in the technical report produced within the research project (in Swedish) [198], and an extended summary is available in a contribution to SFPE's 12th International Performance-Based Codes and Fire Safety Design Methods Conference [199].

Milke et al. [160] has noted that the difficulty with using Jin's work is the association of a very slow walking speed with incapacitation. The same can be said for equating emotional variations to a 4-meter visibility limit as incapacitation. Given the lack of a definitive basis to equate a reduced visibility condition to incapacitation or tenability endpoints, the following considerations are a recommended approach to addressing reduced visibility conditions.

- In large volume spaces or building scenarios using natural or mechanical ventilation schemes that allow the smoke layer to remain at elevations above assumed head height, visibility need not be a factor. To confirm that the smoke layer is adequately above an assumed head height, engineering calculations or fire modeling data should be presented to validate the smoke layer height during the period needed for evacuation or RSET. Consideration with respect to tenability is then needed in terms of the temperature of the smoke layer and radiant flux to occupants beneath the smoke layer.

- Where smoke development and spread is expected to encroach into the path of egress at an assumed head height the reduced visibility condition should be used as a marker or indicator for further analysis. With reduced visibility conditions indicated by engineering calculations or fire modeling data then an additional analysis would:
  - Determine if the area or locality of reduced visibility has completed evacuation prior to the reduced visibility condition.
  - If occupants remain and must egress through the reduced visibility area, then toxic gas/heat exposure must be evaluated to assure occupant tenability for the duration of travel through the reduced visibility area. When occupants find their escape route smoke logged, their willingness to enter the smoke and continue through it depends on a range of behavioral parameters. These parameters include their individual characteristics, their knowledge of the building and escape routes (e.g. familiarity with the space), and their appraisal of the relative risks of seeking refuge or attempting escape. Once exposed to smoke, their willingness and ability to continue also depends upon the optical density, irritancy and temperature of the smoke, ambient/exit lighting, and features of the built environment such as the enclosure size, complexity, and presence of obstacles or other hazards.
  - In those cases, when it is not evident that toxic gas/heat exposure is tenable for occupant evacuation. Then the extent to which smoke density is sufficient to cause a significant proportion of occupants to turn back should be determined. FEC/FED calculations should be conducted to demonstrate that safe egress can be accomplished at reduced travel speeds associated with the reduced visibility conditions. See Table 7.9 for smoke impact on walking speeds:

Table 7.9 Reported effects of smoke on visibility and behavior

Smoke density and irritancy OD/m (extinction coefficient)		Approximate visibility (diffuse illumination)	Reported effects
None		Unaffected	Walking speed 1.2 m/s
0.5 (1.15)	Nonirritant	2 m	Walking speed 0.3 m/s
0.2 (0.5)	Irritant	Reduced	Walking speed 0.3 m/s
0.33 (0.76)	Mixed	3 m approx.	30% people turn back rather than enter

Physical Movement Concepts

8

8.1 Introduction

This chapter addresses the period of time after the decision is made to evacuate or relocate. It provides guidance and quantitative methods for estimating the time for occupants to move to a place of safety or refuge. The guidance describes the quantitative factors and calculation procedures and modeling approaches that can be utilized to estimate the travel or movement time of occupants.

In addition to performing mathematical calculations, the engineer should determine if there is available evacuation drill or case study data and determine if the data are relevant to the context of the building and occupants during a fire incident. Such data may be appropriate as the definitive basis for occupant movement time or may be useful in validating mathematical calculations of movement time.

In the more likely event that no directly relevant movement time data are found, then a suitable calculation method or model needs (as discussed further in Chap. 9) to be selected and used to estimate movement time. Assumptions of the model and the analysis need to be identified as discussed in Chaps. 4 and 6. Where engineering judgment is applied to aspects of the analysis or calculation factors, justification or basis for the engineering judgment should be provided.

Traditionally, the capacity of egress components was measured as a function of the number of units of exit width. (A unit of exit width is defined as 0.56 m (22 in.)). However, the work of Pauls, [200] Fruin, [201] and Habicht/Braaksma [202] has demonstrated that the capacity of egress components for large width components approximates a linear function of the clear width of the egress component, less an edge component along each side of the flow path. The resultant width, referred to as effective egress width, is normally about 0.3 m (12 in.) narrower than the actual clear width measurement [203]. The full flow capacity of egress routes with width (above a minimum required by codes) is now approximated as a direct function of width [204].

The abandonment of the exit unit concept was a major step in recognizing that people do not move in regimented side-by-side lanes down stairs or through doors, but rather move in a staggered arrangement that permits lateral body sway [203]. Measurement and analysis of crowd movement have provided the technical foundation for timed exiting approaches that are used to estimate people movement. Such timed egress approaches are recognized and explained in detail in the *SFPE Handbook of Fire Protection Engineering* [203] in Chap. 59 "Employing the Hydraulic Model in Assessing Emergency Movement". The methods and calculations outlined in the *Handbook* abandon some of the traditional egress assumptions of "units of exit width" and the associated optimistic flow assumptions. Instead, the timed egress analysis is oriented towards evaluating people movement in a soundly based and detailed fashion and considers parameters that include boundary conditions at exits such as walls and handrails, stair geometry, travel speed, and people density. This chapter identifies and reviews how these parameters affect the time of movement of occupants once an evacuation is in progress.

The equations present a single value. The data used to create the calculation methods showed a range of movement speeds at all densities. While the flow equation expresses relationships, actual flows must be expected to vary considerably from the calculated flow. Figure 8.1 shows the average speeds descending stairs reported across a range of studies [205]. The reported speeds are typically measured along the slope of the stairs. There is no apparent trend between the reported average speeds and when the study was conducted or the type of building studied.

© Society of Fire Protection Engineers 2019
SFPE Society of Fire Protection Engineers, *SFPE Guide to Human Behavior in Fire*, https://doi.org/10.1007/978-3-319-94697-9_8

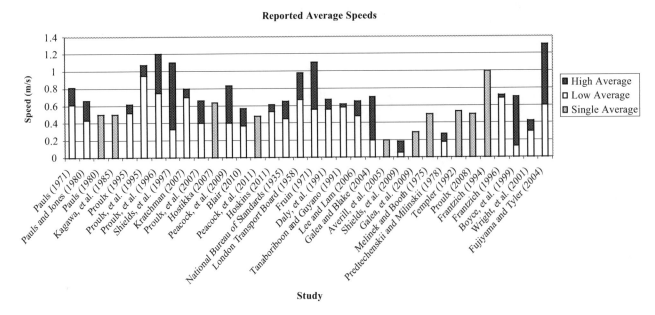

Fig. 8.1 Average Speeds from Studies [205]

These basic calculation methods apply to unimpeded movement along an egress path. However, in some fires, a path is compromised by stored items or fallen debris, emergency responders moving along the same path, etc. Calculation of movement in these situations is sensitive to the nature of the path and the nature of the impediment(s). Guidance and data for engineering calculations of movement in impeded paths can be found in Table 64.17 in the fifth Edition of the *SFPE Handbook of Fire Protection Engineering* [2] and the cited references.

8.2 Factors That Impact Movement Time

The time it takes an occupant to move from his or her starting point to a location of safety is simply a function of travel speed and distance:

$$\text{Time (s)} = \text{Distance (m)}/\text{Speed (m/s)} \qquad (8.1)$$

However, there are many factors that impact both travel speed and distance. Distance is a function of exit choice and exit choice is affected by an occupant's familiarity with the structure, the availability of exits, and the degree of difficulty of an exit path, and other factors as discussed in Chaps. 3 and 4. Some of these factors are occupant characteristics; others are building characteristics. The designer or engineer must either deal with each factor explicitly or be able to justify why a factor is not relevant to the analysis.

There is a range of values that can be calculated to predict the movement component of total evacuation time. The important factors to consider are those that impact RSET. These include one or more of the following: time to clear the building, floor clearing times, stairwell clearing times, time for a person to travel the longest (most remote) path, location of individuals so that exposure levels can be calculated, etc. The calculation method chosen will depend to a great degree on those components of the movement route to which the calculation is most sensitive.

One approach for estimating the evacuation time of occupants applies a hydraulic analogy, simulating people as fluid particles. Another approach considers the behavioral aspects of the people. A prerequisite for either of these approaches is information on the following people movement characteristics:

- Speed: rate of travel along a corridor, ramp, and/or stair (The speed on stairs refers to the rate of travel along a diagonal path obtained by connecting the nosings of the stairs).

8.3 Methods for Calculating Movement Time

- Flow: number of persons passing a particular segment of the egress system per unit time (*e.g.* persons/s passing through a doorway or over an imaginary line drawn across a corridor).
- Specific flow: flow per unit width of the egress component (*e.g.* pers./s-m of doorway width).

Information on people movement has been collected in fire drills and for normal movement. The parameters have been investigated for people movement on stairs, in corridors, and through doorways.

8.3 Methods for Calculating Movement Time

Movement time is calculated based on the density. The movement time can further be used to calculate the flow of occupants past a given point.

8.3.1 Hand Calculations

The *SFPE Handbook of Fire Protection Engineering* in Chap. 59 "Employing the Hydraulic Model in Assessing Emergency Movement" provides a discussion of flow calculations to be used to estimate movement time and the reader is referred to that discussion. [203] It is essential that the engineer keep in mind that these calculations deal only with movement time and can provide non-conservative results. For example, they do not account for scattered evacuation starting times or delays that occur after travel to exits has begun (e.g., the impact of fatigue). Also, they are based on the average movement speed of building evacuations of locations where the majority of the population was working-age adults. Actual evacuation will have people with a range of abilities and some people will not be able to evacuate at those speeds. Those need to be factored in separately by the engineer.

The *Handbook* presents two methods of approximation – both estimate minimum movement time. This chapter discusses the method involving density. The *SFPE Handbook of Fire Protection Engineering* lists approximations for crowd density, speed and flow for various conditions of crowdedness on stairs, along corridors and in doorways. Other approximations can be found in the 2nd edition *SFPE Handbook for Fire Protection Engineering*, Section 3/Chapter 13 under the heading, "Movement Assumptions for Simple, First-Approximation Calculations." The values are based on work by Fruin and Pauls; however they are simplified and optimistic, with no reductions for edge effects or the interactions that occur between individuals while they are evacuating. Pauls [200] reports that using them will result in rough estimates of minimum movement time within 25%. The resultant errors when using these values in calculations of movement times will be acceptable as long as calculated times are considered minimum times for escape movement only. Other behavior, not involving simple movement to the exit, will often be a larger factor in determining total evacuation time. Pauls [200] reports that simple, first-approximation calculations will result in rougher estimates, to within +/− 33%).

Gwynne and Rosenbaum [203] use the methods described above to obtain a first order approximation of the movement time in buildings. The method involves determining the maximum flow rate for each of the egress components in the egress system.

The total movement time is estimated as:

$$t = t_1 + t_2 + t_3 \tag{8.2}$$

where:

t_1: time for first person to reach controlling component.
t_2: time for population to move through controlling component.
t_3: time for last person leaving controlling component to reach place of safety (*i.e.* exterior of building, area of refuge, etc).

8.3.2 Speed

The speed has been shown to be a function of the density of the occupant flow, type of egress component and mobility capabilities of the individual [203].

For a density greater than 0.55 pers./m^2 (0.05 pers./ft^2):

$$\mathbf{v = k - akD} \tag{8.3}$$

For densities less than 0.55 pers./m^2 (0.05 pers./ft^2), based on the Level of Service that Fruin [201] found for when people could freely choose their own speed, too few other people are present to impede the walking speed of an individual. Maximum walking velocities for level walkways and stairways are:

$$\mathbf{v = 0.85k} \tag{8.4}$$

where:
v: speed, m/s (ft/min).
a: constant, 0.266 m^2/pers. (2.86 ft^2/pers).
k: velocity factor, (see Table 8.1), m/s (ft/min).
D: density of occupant flow, pers./m^2 (pers/ft^2).

The values in Table 8.1 for different riser heights and tread depths were not determined experimentally and are approximations. For large riser heights and shorter tread depths speeds will be slower. The selection of the appropriate k value should be based on both the riser height and tread depth. Simply matching one condition (e.g. using $k = 1.08$ when having a 272 mm riser height and 254 mm tread depth) will lead to less reliable results.

At lower densities, people have a greater freedom to move at their own pace. However, their willingness to do so will depend on social bonds (e.g. the people that they are with) and cultural norms (e.g. their reluctance to pass other people). As the crowd density increases, they become more controlled by others in the moving stream. At about 1.9 pers./m^2 (0.18 pers./ft^2), the combination of closeness of individuals and speed of movement are indicated to be the maximum. At this density, exiting individuals would normally be able to see about two treads ahead on stairs or two steps ahead on a flat surface. Such exiting densities are most likely to be encountered in highly populated places of assembly or similar situations involving the movement of significant numbers of persons. This density is not comfortable and it should be expected that, given the opportunity, most persons will increase the space around them, and actually operate at a lower density. Higher densities not only slow the flow but also can reduce movement to a shuffling gait and, in the extreme, a crushing condition.

Figure 8.2 is a plot of Equations 8.3 and 8.4. The speed correlations presented in Equations 8.3 and 8.4 principally relate to average adult, mobile individuals. Proulx [206] indicates that the mean speed on stairs for children under six and the elderly was approximately 0.45 m/s in unannounced drills in multi-story apartment buildings. The speed for an "encumbered" adult is 0.22–0.79 m/s, also appreciably less than the maximum speed noted in Equation 8.4. (An encumbered adult is an individual carrying packages, luggage or a child). Also, slower speeds can be expected for people that require assistance for evacuation. Adams and Galea [207] found that people using an Evac+Chair moved at 0.81 m/s, a carry chair at 0.57 m/s, a stretcher at 0.55 m/s, and a drag mattress at 0.62 m/s. Finally, impaired individuals move slower during an evacuation. Further values for movement speeds can be found in *The SFPE Handbook of Fire Protection Engineering* [2] in Chap. 64 "Engineering Data."

In a more recent study at an assisted living facility, Kuligowski et al. [208] found similar, but slower speeds (average from 0.11 to 0.29 m/s) for people as they descended stairs. In a separate study Kuligowski et al. [209] found an average speed of

Table 8.1 Velocity Factor in Equations 8.3 and 8.4 [203]

Egress Component		k (m/s)	k (ft/min)
Corridor, aisle, ramp, doorway		1.40	275
Stair			
Riser, mm (in)	Tread, mm (in)		
190 (7.5)	254 (10)	1.00	196
272 (7.0)	279 (11)	1.08	212
165 (6.5)	305 (12)	1.16	229
165 (6.5)	330 (13)	1.23	242

8.3 Methods for Calculating Movement Time

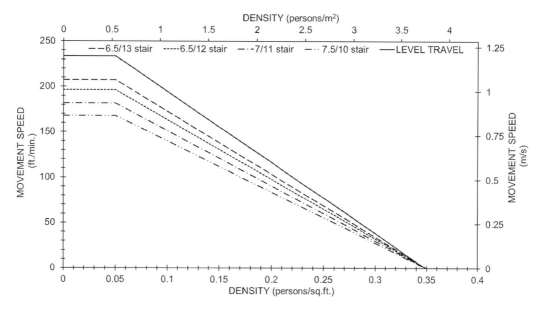

Fig. 8.2 Movement Speed as a Function of Density (Plot of 8.2 and 8.3) [203]

0.31 m/s with a range from 0.07 to 0.91 m/s for older adults in a residential building. Due to the significant spread in the data in all studies, the average values may not be representative of the speed of a given individual.

Care needs to be taken when using the calculation methods. Typically, the values in the calculations are determined based on the average values. Because the goal of the design is typically to ensure that all people remote from ignition are safe, vulnerable sub populations may require more time. The occupant scenarios that are selected must address these populations that are present.

Aside from mobility impairments, other variables have been found to be significant in determining people's movement speeds. While not all studies have found these variables to be significant, examples [1] include age, gender, travel distance, pre-evacuation time, positioning relative to other people, aggressiveness, and encumbrances.

Typical densities of people movement range from 1.0 to 2.0 pers./m^2 (0.09–0.19 pers./ft^2) [201, 203, 210, 211] While the equations can be solved for higher densities, these values were typically not part of the original data set and extrapolation to those values will go beyond the validity of the equations. Also, for calculating the density on stairs, the typical approach has been to use the plan view to find the area. Hoskins [212] has developed a method to compare the densities based on the adjusted tread depth that the person uses (approximately 230 mm) [204]. The density of a flow can be determined in the following ways:

- The ratio of the number of people in a group in an egress component divided by the total floor area occupied by the group (including the area between individuals). For the equations in this guide, the density is defined in terms of this approach.
- The ratio of the floor area occupied by each individual person in the group divided by the total floor area occupied by the group (including the area between individuals).

8.3.3 Specific Flow

By combining the concept of people movement speed with effective width, the number of persons moving past a point in the egress route per unit time per unit distance of effective width can be determined. This flow is termed "Specific Flow" and is calculated as the speed multiplied by the density of the population.

The specific flow is analogous to the mass flux in hydraulic systems. As such, the specific flow is the product of the density and the speed:

$$\mathbf{F_s = Dv} \tag{8.5}$$

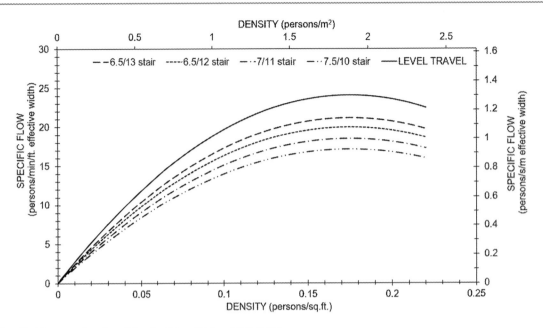

Fig. 8.3 Specific Flow as a function of density (Plot of 8.6 and 8.7) [203]

Table 8.2 Boundary Layer Width [203]

Component	Boundary Layer, mm (in)
Theater chairs, stadium benches	0 (0)
Railings, handrails[a]	89 (3.5)
Obstacles	100 (4)
Stairways, doors[b], archways	150 (6)
Corridor and ramp walls	200 (8)

[a]Where handrails are present, use the value that results in the lesser effective width
[b]There are no data to substantiate the effective width of doors; the value is an extrapolation from the value used for stairs, and is believed to result in estimates that do not under estimate the movement time through doors

Expressions for the specific flow as a function of density only can be obtained by substituting for the speed from Equations 8.3 and 8.4, these are shown in Fig. 8.3:

For a density greater than 0.54 pers./m^2 (0.05 pers./ft^2):

$$\mathbf{F_s = Dv} = (1 - \mathbf{aD})\mathbf{kD} \tag{8.6}$$

For densities less than 0.54 pers./m^2 (0.05 pers./ft^2):

$$\mathbf{F_s} = 0.85\mathbf{kD} \tag{8.7}$$

where:

F_s: specific flow, pers./s-m (pers/min-ft).

As discussed is Sect. 8.1, the width referenced in the units for the specific flow equations relates to the "effective width" as defined by Pauls [200]. The concept of effective width is based on the observation that people do not generally occupy the entire width of an egress component, staying a small distance away from the walls or edge of the component. Gwynne and Rosenbaum [203] refer to this small distance as a "boundary layer," in keeping with the hydraulic analogy for people movement. The width of the boundary layer for the variety of egress components is presented in Table 8.2.

8.3 Methods for Calculating Movement Time

Table 8.3 Maximum Specific Flows [203]

Egress Component		F_s pers./s-m of effective width (pers/min-ft of effective width)
Corridor, aisle, ramp, doorway		1.32 (24.0)
Riser, mm (in)	Tread mm (in)	
190 (7.5)	254 (10)	0.94 (17.1)
272 (7.0)	279 (11)	1.01 (18.5)
165 (6.5)	305 (12)	1.09 (20.0)
165 (6.5)	330 (13)	1.16 (21.2)

Considering the quadratic function of the specific flow, a maximum specific flow is achieved at a density of:

$$\mathbf{D_{max}} = 1/2\mathbf{a} \tag{8.8}$$

Because a is independent of the type of egress component, according to this correlation the specific flow is maximized at the same density for all types of egress components. Predtechenskii and Milinskii [203] provide results from their data indicating differences in the density where the specific flow is maximized for different types of egress components.

In understanding the concept of "Specific Flow," it is important to realize that the flow of adults can reach only a specific maximum value as illustrated in Fig. 8.3. For children, the density and specific flows are greater than the typical values for adults [214]. By reviewing Fig. 8.3, starting at a density of 0 and speed of 0, it is noted that the "Specific Flow" of occupants is shown to increase as the density increases. Although the speed of moving occupants is decreasing, it is more than compensated for by the greater density of moving people. However, the "Specific Flow" begins to diminish and an optimum flow condition is reached at a density of approximately 1.94 pers./m^2 (0.18 pers./ft^2 or 1 pers./5.6ft^2). At densities greater than 1.94 pers./m^2, their speed continues to decrease, but the increasing crowd density rather than compensating for the continuing decreasing speed now makes occupant flow more difficult until the "Specific Flow" would theoretically drops to 0 (these densities were not observed in previous studies).

Recognizing that flows may vary significantly between optimal and less than optimal conditions, the timed exiting evaluation of a building or portion of a building may need to consider a range of conditions rather than being based solely on optimal flow conditions.

The maximum flow rate occurs when the specific flow is maximized (i.e. where D_{max} occurs, see Equation 8.8). Maximum specific flows for a variety of egress components are provided in Table 8.3. The controlling egress component is the component with the smallest maximum flow rate, relating to where a queue is expected to form if D_{max} occurs in an upstream component. It is typically assumed that if there are enough people present, the maximum specific flow will be attained. However, such an assumption is an optimization, and analyses should account for uncertainty in this assumption.

8.3.4 Total Flow Capacity

The "Specific Flow" mentioned above provides a measurement of the flow capability of an egress component on a per unit width basis (e.g., per meter). In the evaluation of an egress component or multiple egress components, the total flow can be calculated and related to the affected population. Multiplying the "Specific Flow" by the total effective width of all exits permits the calculation of a predicted flow rate of persons passing through an exit route or routes. Hence,

$$\mathbf{Flow\ Capacity} = \mathbf{F_c} = \mathbf{F_s W_e} \tag{8.9}$$

$$\boldsymbol{Substituting\ F_s} = (1 - aD)kD \tag{8.10}$$

$$\boldsymbol{Yields\ F_c} = (1 - aD)kDW_e \tag{8.11}$$

The movement time for a populated area through one exit element is the population, P, divided by the flow capacity of the exit element, plus the travel time through the exit element. The following examples demonstrate the use of this method.

8.4 Examples

8.4.1 Example #1

The total movement time for a room containing 300 people will be determined. The room has a maximum travel distance of 60.96 m (200 ft) to egress through two 0.81 m (32 in.) doors that lead to two enclosed 1.12 m (44 in.) stairs (height and depth of stair tread, determined to be 272 mm (7 in.) and 279 mm (11 in.), respectively) and down 15.24 m (50 ft) of stairs to a wide discharge at grade.

Assuming that neither of the stair entrance doors are blocked by the fire and that the occupants in the room are equally distributed (low density; 0.54 pers./m² (0.05 pers./ft²)), one would first consider if the time to travel to the stairs is greater than the time for occupants to move through the doors into the stairs or move down the stairs.

$$\mathbf{T} = \frac{\mathbf{d}}{\mathbf{v}} = \frac{\mathbf{d}}{0.85\mathbf{k}} = \frac{60.96 \text{ m}}{0.85(1.4 \text{ m/s})} = 51 \text{ s}$$

The time for occupants to move through the door or on the stairs will indicate if occupants are queuing at the stair entry doors.

Movement time through the doors:

$$\mathbf{T} = \frac{\mathbf{P}}{F_{S_{max}} W_e} = \frac{300}{(1.32 \text{ pers/s} - \text{m})(2\,[0.81 \text{ m} - 2(0.15 \text{ m})])} = 223 \text{ s}$$

Movement time on the stairs:

$$\mathbf{T} = \frac{\mathbf{P}}{F_{S_{max}} W_e} = \frac{300}{(1.01 \text{ pers/s} - \text{m})(2\,[1.12 \text{ m} - 2(0.15\text{m})])} = 181 \text{ s}$$

Since the movement time on the stairs is less than the movement time through the doors, the stair entry doors control the flow of occupants. All of the occupants have moved through the stair entry doors in approximately 223 s (3.8 min) (assuming that their density is D_{max}), while it only takes 51 s (0.9 min) and 181 s (3.0 min) for the occupants to travel to the doors and past a point on the stairs.

Movement time down the stairs:

$$\mathbf{T} = \frac{\mathbf{d}}{\mathbf{v}} = \frac{\mathbf{d}}{0.85\mathbf{k}} = \frac{15.24 \text{ m}}{0.85(1.08 \text{ m/s})} = 17 \text{ s}$$

The total movement time, assuming once again that the occupants in the room are equally distributed and the first occupants to egress are initially at the stair entry doors, equals 223 s (3.8 min) for moving through the stair entry doors plus 17 s (0.3 min) for the last person to walk down the stairs. This results in a total movement time of approximately 240 s (4.1 min).

NOTE: It is important for the user to analyze each individual situation carefully, considering the realm of possible building and occupant arrangements. For instance, in the above example, one could assume that one of the stair entry doors was blocked by the fire or that all occupants were in a meeting in the far corner of the space. Adjustments to the calculations would then be made, resulting in different and increased total people movement times.

8.4.2 Example #2

Determine the total movement time for a 5-story building with the following characteristics:

There are 200 people on each floor. Each floor is served by two 1.12 m (44 in.) wide stairways. The doors leading into and from the stairway are 0.81 m wide (32 in.). The stair design includes 178 mm (7 in.) risers and 279 mm (11 in.) treads. The floor-to-floor distance is 3.66 m (12 ft) and the landing between floors is 1.22×2.44 m (4×8 ft). Handrails are provided on both sides of the stairways.

Solution:

Component	Effective Width, m (ft)	Specific Flow, pers./m-sec (pers/ft-min)	Flow, pers./sec (pers/min)
Door into stairway	0.51 (1.67)	1.31 (24.0)	0.67 (40)
Stairway	0.94 (3.08)	1.01 (18.5)	0.95 (57)
Landing	0.82 (2.67)	1.31 (24.0)	1.1 (65)
Door from stairway	0.51 (1.67)	1.31 (24.0)	0.67 (40)

The controlling component is the door leading from the stairway. The time required for the half of the building occupants on the upper floors (400 persons) to pass through this doorway is estimated as.

$$T = \frac{P}{F_{S_{max}} W_e} = \frac{400}{(1.32 \text{ pers/s} - \text{m})(0.81 \text{ m} - 2(0.15 \text{ m}))} = 594 \text{ s}$$

Next, calculate the time required for the first person to travel down the stairs.

Time to travel down one flight of stairs:

The hypotenuse of 178/279 mm (7/11 in. stair is 0.33 m (13 in). Thus, to travel a vertical distance of 3.66 m (12 ft) requires traveling a diagonal distance of 6.8 m (22.3 ft). Using eq. 5.1, the speed at D_{max} of 1.88 pers./m^2 (0.18 pers./ft^2) is 0.54 m/s (106 ft./min).

For travel distance on the landing, two different methods have been proposed. The first is to use twice the stair width based on the midpoint path using straight lines. However, the work of Templer [215] found that people followed arcs rather than straight-line paths. Based on this and his own observations of evacuation drills, Hoskins and Milke [205] proposed that a more accurate travel distance on landings can be determined based on Equation 8.12.

$$\textbf{Landing} = \boldsymbol{\pi}\textbf{W}/2 \tag{8.12}$$

where: W = the width of the stair.

The length of travel along each landing is 1.9 m (6.3 ft) (assuming an average length of travel on the arc through middle of the landing). Because the speed on a stairway is less than that for a horizontal component such as a landing, the speed on the landing is limited to that achieved on the stairway. As such, the length of travel on the landing can be added to that for the stairway, giving a total length of travel of 10.6 m (34.9 ft). The time required to traverse this distance at the speed achieved on the stairways is 19 s (0.31 min). Since a queue is expected to form at the door leading from the stair, it is only necessary to consider the time for the first person on the second floor to travel down one flight of stairs to the first floor, as people coming from upper floors are expected to encounter a queue. As the time that people would wait in the queue would be greater than the time that would be needed to travel down the stairs if the queue did not exist, we can neglect the time that it would take for people on upper floors to travel down the stairs. This results in a total movement time of 19 s + 594 s = 613 s (10.2 min), neglecting the time needed to travel to the stairs.

This type of analysis is most relevant in situations where a queue is expected to form at the controlling egress component. Generally, these situations consist of cases where an appreciable number of people occupy the area of the building being modeled. Conversely, in buildings with low occupant loads, a queue is unlikely. In cases with low occupant loads, a more complex analysis is needed to examine the occupant flow on a component by component basis. These analyses also may be applied to provide a more accurate assessment in cases where queuing is likely.

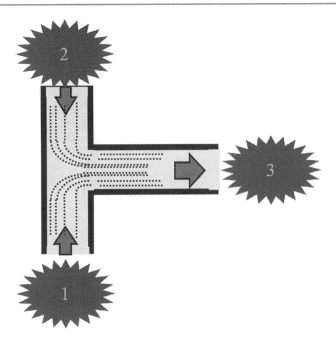

Fig. 8.4 Merging Egress Flows [1]

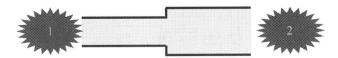

Fig. 8.5 Transition in Egress Component [1]

The component-by-component analysis involves a determination of the time for the population to traverse each egress component. Transitions between different components, changes in component width and mergers are addressed via the following rules:

- The combined flow rate of people entering an intersection equals the flow rate of people from the intersection (see Fig. 8.4):

$$\mathbf{F}_{C1} + \mathbf{F}_{C2} = \mathbf{F}_{C3} \qquad (8.13)$$

If the combined flow rate of egress components leading to the intersection is greater than the capacity of the flow rate for the egress component leading from the intersection, a queue is expected to form. If a queue forms, the analysis can continue, considering that the flow rate in component #3 is equal to the maximum capacity of the component.

→ Where the width of the egress component changes, then the density of the flow is also expected to change. The new density is determined by the following relationship:

$$F_{C1} = F_{C2} \qquad (8.14)$$

Again, if the incoming flow rate leading to the transition point is greater than the capacity of the flow rate for the egress component leading from the transition, a queue is expected to form at the transition (Fig. 8.5).

The density of a flow of occupants proceeding away from a transition is determined by solving either Equation 8.6 or Equation 8.7. Where Equation 8.6 applies, solution of the quadratic equation results in two possible solutions for the density. The lesser value for density should be selected as the correct value. The lower density is correct, since if an occupant flow at the maximum density was approaching a widening corridor, the solution of Equation 8.6 would yield one density greater than the maximum and one less. However, in the case of a widening corridor, it's unreasonable to expect the density to increase (and speed to decrease) from the narrow to the wide corridor.

8.4 Examples

In either of these types of analyses where multiple egress paths are available to a group of occupants, some assumptions need to be made of where the occupants are located. Often, an equal proportion of the group is assumed in each of the available paths. Alternatively, the distribution may be determined in proportion to the respective capacities or other characteristics of the available paths [213, 216].

Gwynne and Rosenbaum [203] recommend the following method of determining the densities and flow rates following the passage of a transition point:

1. The flow after a transition point is a function, within limits, of the flow(s) entering the transition point.
2. The calculated flow, F_c, following a transition point cannot exceed the maximum specific flow, F_{sm}, for the route element involved multiplied by the effective width, W_e, of that element.
3. Within the limits of rule 2, the specific flow, F_s, of the route departing from a transition point is determined by the following equation:

   (a) For cases involving one flow into and one flow out of a transition point:

$$F_{s(out)} = F_{s(in)} W_{e(in)} / W_{e(out)} \tag{8.15a}$$

   where:
   $F_{s(out)}$ = specific flow departing from transition point,
   $F_{s(in)}$ = specific flow arriving at transition point,
   $W_{e(in)}$ = effective width prior to transition point, and
   $W_{e(out)}$ = effective width after passing transition point.

   (b) For cases involving two incoming flows and one outflow from a transition point, such as that which occurs with the merger of a flow down a stair and the entering flow at a floor:

$$F_{s(out)} = \left\{ \left[F_{s(in-1)} W_{e(in-1)} \right] + \left[F_{s(in-2)} W_{e(in-2)} \right] \right\} / W_{e(out)} \tag{8.15b}$$

   where:
   the subscripts *(in-1)* and *(in-2)* indicate the values for the two incoming flows.

   (c) For cases involving other merger geometries the following general relationship applies:

$$\left[F_{s(in-1)} W_{e(in-1)} \right] + \ldots + \left[F_{s(in-n)} W_{e(in-n)} \right] = \left[F_{s(out-1)} W_{e(out-1)} \right] + \ldots + \left[F_{s(out-n)} W_{e(out-n)} \right] \tag{8.15c}$$

   where:
   the letter n in the subscripts *(in-n)* and *(out-n)* is a number equal to the total number of routes entering *(in-n)* or leaving *(in-n)* the transition point.

4. Where the calculated specific flow, F_s, for the route(s) leaving a transition point, as derived from the equations in rule 3, exceeds the maximum specific flow, F_{sm}, a queue will form at the incoming side of the transition point. The number of persons in the queue will grow at a rate equal to the calculated flow, F_c, in the arriving route minus the calculated flow leaving the route through the transition point.

5. Where the calculated outgoing specific flow, $F_{s(out)}$, is less than the maximum specific flow, F_{sm}, for that route(s), there may be no way to predetermine how the incoming routes will merge. The routes may share access through the transition point equally, or there may be a total dominance of one route over the other. For conservative calculations, assume that the route of interest is dominated by the other route(s). If all routes are of concern, it is necessary to conduct a series of calculations to establish the bounds on each route under each condition of dominance.

Egress Model Selection

9

9.1 Introduction

The rapid increase in computer capability and decrease in associated cost have expanded the use of computer models in all fields of engineering. This has particularly been the case for simulation tools, such as egress models. An engineer performing a life safety analysis on a structure is presented with several alternative tools from which to choose to complete this task. Depending on the type of building and time and budget constraints, the engineer may choose from a variety of techniques, including empirical calculations, manual engineering calculations, and/or computational simulation modeling.

The empirical engineering approach compares the structure in question to data collected from comparable structures. The engineer can then extrapolate from those data in order to make a prediction of the egress performance of the current structure of interest. This type of analysis and the results produced operate at a high level, looking at overall evacuation times, rather than the performance of individual elements in the structure. For that reason, this approach is now less commonly used, as more refined insights into the outcome of an evacuation are now typically required.

Manual engineering calculations applies empirical data at the component level (doorways, stairs, etc.) to ascertain the egress performance across the structure once these components are chained together into an egress path. This is done by predicting the time taken for evacuees to move within the components in question along predefined routes within the structure. This assumes people move collectively as a (hydraulic) flow along a component. This flow is determined by the component's capacity (i.e. the space available), type (e.g. stairs or walkways) and the expected physical performance of the evacuees.

Manual engineering calculations, as part of the hydraulic approach, are commonly used by engineers to estimate egress times. The reasons behind these calculations being favored are likely tied to many factors: i.e., existing levels of expertise (i.e. engineers are familiar with it), model availability, project time/budget constraints, and the availability of data. While this approach is not inherently incorrect, it does involve numerous simplifications and the exclusion of several important factors, requiring user efforts to compensate for this exclusion (e.g. using safety factors). Engineering calculations are covered in detail in Chap. 8. It is essential that the engineer keep in mind that such hand-based engineering calculations only deal with *physical elements* of egress – i.e., movement time - and may not provide optimal results. For example, by default, they do not account for evacuation starting time delays that occur during a building evacuation, the distribution of evacuees between exits, variability in evacuee travel speeds, etc. Overall, when using this approach, the user is required to account for decisions and behaviors of the population separately.

Finally, computer simulation models represent a more diverse set of methods and sophistication, ranging from homogeneous occupant flow (using hydraulic or other flow methods), to autonomous agents moving throughout three-dimensional space. There are significant differences in methodology between current computer simulation models, however, they have one distinct element in common: their predictions are based on a set of initial conditions that, over time, provide insights of interest. This differentiates them from the empirical and engineering calculations which are less detailed in nature. This section identifies a set of questions and issues that need to be addressed by the egress model user to determine his/her confidence in selecting one or both of these last two options (i.e., engineering calculations or computer simulation models).

Unfortunately, there is no one-size-fits-all guidance that clearly determines when a specific model or type of model should be selected. The following discussion provides the user with a set of tools to inform this decision, including factors to consider and questions to ask of the models. In the recent past, the development and use of computational egress models has grown due

© Society of Fire Protection Engineers 2019
SFPE Society of Fire Protection Engineers, *SFPE Guide to Human Behavior in Fire*, https://doi.org/10.1007/978-3-319-94697-9_9

to technological developments as well as the demand for flexible techniques that can cope with complex designs. In addition, the needs of the fire protection/safety community to address problems other than building fires have been apparent; for instance, egress models have been used to simulate evacuation from public disorder incidents, bomb threats, active-shooter events, hazardous weather, and even relocation (instead of evacuation) within the building. Egress models have also been used to simulate non-evacuation events such as building ingress and routine occupant movements. This chapter discusses some of the factors that may aid an engineer in selecting a model.

Before selecting a model, an engineer should identify scenarios of interest and decide what measures need to be taken in order to quantify and compare conditions for safety. The scenarios of interest are typically selected to answer questions of the design and safety provisions that provide some insight into real-world conditions. Ideally, the egress model will capture all relevant aspects of the situation being examined in a reliable and credible manner; e.g. the evacuee behavior, the conditions faced, etc. However, this is never the case. The model used, by definition, is a simplification that excludes elements of the scenario of interest. This simplification is often exaggerated given the time, expertise and resources available; i.e. the most relevant, capable, credible model may not always be the one that is available or selected for various reasons (see Chap. 10).

It is important that the engineer understands the extent of this simplification for two reasons: (1) to potentially compensate by configuring the model manually ('driving the evacuee response' to better represent the scenario), and (2) to place the credibility of the modeled results in context for third parties (see Chap. 11). To understand the gap between the model assumptions/capabilities and the scenario the user must have an understanding of the real-world factors being examined (see Chaps. 4, 5, 8 and 13) and then determine what the model is able to do and what can be done to compensate for the discrepancies between the model and the scenarios of interest. This involves expertise in the modeling and subject domains (i.e. engineering and the social sciences). This chapter will also help the engineer identify the questions that need to be asked of the model to establish the extent of the gap between real world scenarios of interest and the model capabilities and outcomes. Asking these questions will inform the engineer's ability to document and minimize this gap – either through model selection or action to calibrate and compensate for underrepresented facets of the scenario.

9.2 Project Considerations

In order to select the appropriate egress model, it is important that the user consider key questions relating to the suitability of the model to the project at hand. Kuligowski and Gwynne [217] suggest the user answer these questions to ascertain which model is appropriate for the project:

1. What information is needed to frame the scenario and configure the model? Ideally data derived from the structure being examined during a representative scenario would be available. However, for new buildings/ships/other structures, the egress models are typically employed prior to final design, precluding the availability of such data. Instead, the engineer will likely have to search for data from comparable structures and situations to configure the model being used. This can be a challenge – and can preclude the use of some models that are particularly information hungry. At the very least the absence of relevant data can reduce the benefits of employing more advanced (and potentially more expensive) models. This leads to the questions: *What is the engineering question I'm trying to answer? What factors drive that answer? What information is needed to represent these factors within the models available and what information is available?*

2. How much time and funding are available to complete the project? The reality within the fire engineering community is that there are limitations to what can be accomplished based on the resources available. The application time and cost of a model is a factor in model selection. Recognizing that cost is a factor leads to the following questions: *What level of effort is required to configure and execute the model? What is the nature and scope of the project? What are the conditions that need to be represented? What are the technical resources required? What additional human resources (training, expertise, etc.) are required?*

3. What is the practitioner's level of expertise and familiarity with the candidate modeling tools? If insufficient expertise is available for a particular model then additional time will be required to ensure training and research is conducted. It would be unusual for training/research to be budgeted inside a typical modeling project. It is more likely for practitioners to default to a model with which they are familiar and then enhance their expertise to better address the scenarios of interest. Example questions to ask include: *How familiar is the practitioner with the candidate models? How familiar is the practitioner with this type of application and the model configuration required?*

4. What is the practitioner's level of expertise and familiarity with the evacuation dynamics associated with the scenarios of interest; for instance, the expected conditions and evacuee behaviors? Example questions: *Does the practitioner have*

9.3 Model Attributes

sufficient understanding of the expected scenario conditions and population response to determine what model capabilities are required and any shortfalls that might exist? Is there sufficient time for any gaps in subject matter expertise to be addressed within the project – either through training or new resources?

5. What are the deliverables of the project? Where visual output is required to satisfy an Authority Having Jurisdiction (AHJ), a model that produces text output might not suffice for approval. As outlined in the *SFPE Engineering Guide to Performance Based Design* [3], all stakeholders should be involved with the decision-making process related to what project outputs are required. This prompts the following question: *What output needs to be produced by the model to meet the project deliverables?*

6. What is the purpose of the project? The purpose will determine the scope of the factors to be modeled, the content/format of the results required, the analysis conducted and the way the results are communicated to the client. Models have a number of applications that include:

   - Determining the Required Safe Egress Time (RSET) as part of a performance-based design (i.e. the time to evacuate or relocate occupants within a space or building to a place of safety). This is the most common application and typically also requires the determination of the Available Safe Egress Time (the time for conditions to become untenable) for comparison.
   - Determining allowable occupant load based on relocation time to an adjacent space.
   - Determining whether additional egress resources are required (e.g. following changes to the use of a space, etc.). This may consider existing occupant loads that exceed exit capacities for existing buildings, or changes to internal arrangements that cause a change in occupant load beyond the designed exit condition.
   - Comparative analysis to the prescriptive requirements.
   - Establishing the cost effectiveness of design changes.
   - Procedural design changes to existing buildings.
   - Forensic analysis.
   - Research Applications
   - Evacuation Demonstration – as part of occupant or responder training.

7. Does the project warrant or enable the application of a computer simulation? Are other options available if: (1) insufficient data/information is available to configure a computational model and exploit its relative 'sophistication'; (2) the proposed scenario of interest is so simple (e.g. direct movement to a place of safety, relatively uniform population, low population density and few evacuee/flow interactions), that the outcome can be calculated directly by a trained practitioner; and (3) the computer models available have insufficient validation in the scenarios being suggested such that the credibility of their projections are questionable (see Chap. 10). The application of safety factors at each stage of the analysis may be suggested to compensate for some of these (and other) issues. Although necessary, these may reduce the capacity of the models to discriminate between different designs/scenarios – a key benefit of such models (see Chap. 11).

9.3 Model Attributes

The next step in the modeling process is identifying candidate models – based on their attributes and capabilities. This task is often governed by matters of availability, expediency, and economics rather than based on selecting the model best suited for the task. Of primary importance in the selection process is an understanding of the background of the models and their current characteristics (i.e., both capabilities and limitations – see Fig. 9.1). By understanding this, the user can differentiate among the models available and make a more educated choice. For many models, information can be found on the developers and the model assumptions from the model users' guide and associated documents. If this information is not available then it is a significant problem. This lack of information may, in and of itself, preclude a model from consideration.

An important aspect of model selection is determining the level to which the model has been subjected to validation. This is discussed in detail in Chap. 10. Validated models are those found to be supported by comprehensive and available testing documentation. The absence of such documentation poses issues of credibility for the user. Egress models rely on observed research data, mathematical algorithms, and testing at the level of refinement at which the model operates; i.e. at the level that will influence expected model performance. Therefore, refined computational tools have more opportunity and more requirements for testing giving the additional levels of detail at which they might operate. In contrast, manual engineering

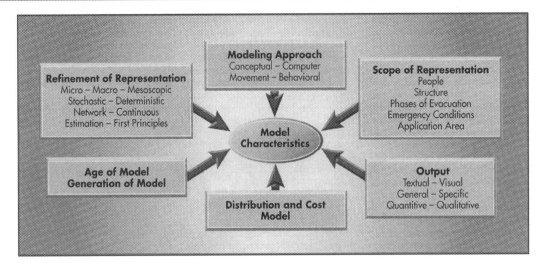

Fig. 9.1 Model Characteristics and Model Selection Relationship [217]

calculations adopt a higher-level approach where people movement is dictated by fluid flow calculations, with changes in speed based on density and route capacity reducing the levels of refinement at which comparisons can be made.

Figure 9.1 provides an outline as to how the model characteristics and model selection are affected by the needs of the analysis. This will be used to structure the discussion below.

The model characteristics affect the type of scenarios that can be produced and the manner in which the scenarios will be developed and will therefore be further discussed below in the context of scenario development.

9.4 Developing Model Scenarios

Once the user is familiar with the project requirements and model features, the user should begin developing possible scenarios for model configuration. The purpose of this section is to outline the choices a user will have to consider and offer guidance on how to make the decisions that will eventually lead to application of the model. The user should refer to Chap. 6 for further information on scenario development.

As with model testing, the more refined the model (e.g. a behavioral model that represents a number of evacuee actions at the individual level), the more data is likely to be required to configure the model for application. This may take time and expertise and rely on the existence of data to complete the task. It should also be remembered that the level of detail at which the model operates does not necessarily translate into more accurate output. It provides the *opportunity* to represent certain elements in more detail; but does not ensure more credible results. This has to be confirmed through validation (see Chap. 10).

> Engineers conducted a behavior study and modeling exercise effort [218] to ascertain the impact of installing revolving security doors at the entrance of a secure building on occupant congestion. As a first order approximation, hand calculations were used to determine if congestion would occur by simply comparing the average occupant ingress rate with the total throughput capacity of the revolving doors (given the best research evidence available). Computer modeling was not employed at this stage, given the simplicity of the situation being examined.
>
> A more detailed analysis was then conducted, using field observations (as the building was already constructed), where the revolving door throughput rates were measured under actual building conditions. Observations demonstrated that the flow was varied, due to arrival patterns and occupant badging into the structure. These flow dynamics were deemed too complex to characterize within hand calculations and so a computer model was employed that was capable of generating fluctuating flow patterns based on individual actions; i.e. to capture the observed conditions.
>
> The calibrated modeling analysis showed that the initial first order approximation, using constant door throughput rates and occupant ingress rates, resulted in overly-optimistic results and did not predict periods of occupant congestion that were observed in the field during population bursts.

9.4 Developing Model Scenarios

9.4.1 Building Configuration (the "Structure")

When using a model, the user must describe the building characteristics in a manner consistent with the method used by the model to represent the geometry of the building. This will depend on the level of refinement at which the model represents the space and also the functionality offered by the model.

This description includes the location of open spaces and walls, information on the stairs, and the location of the final destination of safety. The user may be required to provide these data in a number of different ways. The modeling method employed to represent the structure will directly influence the results produced; therefore, the user should seek information within the user's guide relating to the method being applied.

There are several methods by which a model might represent the structure. If a *coarse network* is employed within the model, the model represents the floor plans of the structure as a segmented version of the building. This will normally relate to the connectedness and capacity of each of the structural components (e.g., a room, a corridor, a stair, etc.). When employing a *fine grid structure*, the model overlays a mesh of small uniform nodes throughout the occupiable space within the entire floor area, with each node typically representing an estimate of the space occupied by one person. These nodes connect to each other to form the locations and then paths that occupants might use to travel *between and within* the structural components. Several models employ a *continuous representation* of the space that produce a representative continuous plane for each level of the occupiable space within which the occupants move across a coordinate system. Some models allow the use of CAD files (particularly important for continuous models) or Building Information Model (BIM) files to describe the spatial configuration. This will influence the time that it takes the user to represent the structure within the model. Several questions should be asked of candidate models:

- *Does the structural complexity require the use of CAD to capture its design?*
- *Given the population size, does the structural capacity affect evacuee performance?*
- *How does the model allow for structure generation?*
- *How does the model allow for structure representation?*
- *What is involved in configuring the representation of building?*
- *What other kinds of information are needed to configure the representation of the building?*

9.4.2 Population Configuration (the "People")

The building needs to be populated before the simulation can proceed. In order to do this, the following information is required (at a minimum): the number of occupants and their distribution throughout the building, their characteristics that can affect their performance within the model (e.g., age, gender, knowledge level, exit familiarity, impairment, etc.), data to describe their movement abilities (e.g. achievable speeds, flow/density relationship, etc.), pre-evacuation delays, other specified behaviors and associated delays (for instance, whether the evacuees have tasks to complete prior to evacuation, have responsibilities during the evacuation process, etc.). More guidance on this can be found in Chaps. 3 and 6.

As the model's sophistication increases, the number of behavioral variables considered may also increase, potentially increasing the number of possible input parameters. It is then up to the user to select input parameters that specify the entire population, subpopulations, or individuals throughout the building. Often the models provide default values; however, the user should never blindly adopt default values - but instead be informed about the assumptions on which they are based and the scenarios to which they might apply. Only then can users make informed decisions about whether they pertain to the population involved in the project at hand.

A movement study was conducted at a large airport [219] in order to perform field observations for use in computer simulation analysis. Occupant movement characteristics and grouping behaviors were observed and recorded.

The flow through doors with roller bags was observed to be less than expected for unencumbered movement; in addition, the exit capacity was deemed to be less when bags were present. The impact of these findings on airport performance was demonstrated by simulating the performance where none of the population had rolling bags and where the observed proportion of the population had rolling luggage. The simulation work was then able to determine how optimistic the previous assumptions might have been and, in turn, more credible performance levels.

There are two fundamental methods to represent the evacuating population within the structure. The simulated population can be represented at the *micro–level*, as individuals with local characteristics, or at the *macro-level* as a homogeneous group given population-wide characteristics. The former approach is employed by several computational simulations; this latter approach is employed by the hand-based (or manual) engineering calculations, as well as by several simpler computational approaches. The following questions should be posed of candidate models:

- *Is the model able to represent the variability in the attributes and actions of the population?*
- *How many occupants are in the building?*
- *What movement data are required by the model?*
- *What occupant characteristics are required by the model?*
- *What pre-evacuation data are required by the model?*
- *What kinds of behavioral inputs are of interest for the population and is there information available to provide as input?*

9.4.3 Procedural Configuration

These procedures relate to expected situations and protocols within the scenarios of interest. These may be formal emergency procedures employed within the structure to manage the evacuation response (e.g. staged evacuation plan) or factors that influence individual response selection (e.g. familiarity influencing the route choice). The user has to choose the evacuation procedures and situations included to complete particular scenarios, assuming that the model is capable of doing so. These include:

- The route choice and exit knowledge of the occupants
- The existence of emergency responders and building staff (potentially producing counter-flow)
- The inclusion and activity of human and technological resources that influence evacuee response (notification, signage systems, etc.).

There may also be non-fire emergency procedures that have an influence upon performance during an evacuation; for instance, security constraints, routine use of the building, location of facilities, etc. These may have profound influences over the evacuation results. For instance, routine use will dictate the initial conditions of an emergency; security considerations will influence the routes with which people are familiar and their likelihood of using certain routes, etc. These also represent simulation applications in their own right – presenting the model user with interface issues (e.g. can the model represent the transition between non-emergency and emergency conditions?) and scope issues (e.g. can the model address the interaction between security and evacuation procedures?). Questions to be asked of the model include:

- *Is the model able to represent procedural aspects present within the scenarios of interest that will influence evacuee performance?*
- *Is there an emergency procedure for the building or for parts of it?*
- *What type of notification and signage system in place?*
- *Is exit choice based on proximity, familiarity, or some combination of factors within the model?*
- *Is the model able to represent counter-flow produced by the evacuating population moving against fire fighters or building staff?*

9.4.4 Environmental Configuration

The user has to decide whether the fire is represented at all, implicitly or explicitly. If the fire conditions are not considered, then no further actions are required, although this fact should be clearly reported. Within these models (where fire conditions are not considered), the user would have to manually determine when fire conditions reached a level that would influence the behavior and well-being of the evacuating population by examining external information (e.g. manually examining the results of a third-party model).

9.4 Developing Model Scenarios

If fire conditions are considered implicitly, then the fire will not appear within the evacuation model, but may instead act as a benchmark for comparison after the results are produced. If the fire is considered explicitly, then depending on the nature of the model, the effect of the fire might be supplied manually by the user or represented directly within the simulated environment using fire predictions (e.g. from a zone model or CFD model).

Although there may be many ways in which the environment influences evacuee performance, models typically simplify the impact given that the evacuating population is expected to move to minimize their exposure. The primary impact of fire conditions on the evacuation is typically via attainable travel speeds, the routes available, and the physical condition of the evacuating population.

As mentioned, a model can either include the impact of fire conditions on the evacuation or require the user to make comparisons between the conditions manually to draw conclusions. This comparison is common within the fire protection community for ASET/RSET analyses. An example of a manual comparison of two models is shown in Fig. 9.2. The locations of failure (i.e. the prediction of untenable conditions) are represented and can be compared to the location of occupants based on the behavioral computer-based model analysis.

Automatic overlaying of model results (coupled within the same simulated environment), incorporate the findings of the fire model into the egress model. Figure 9.3 shows a representation of an automatic overlaying of two models.

Coupled models are two (or more) models that produce results from different domains that interact within the same modeling environment. One-way coupling is when multiple sets of output exist in the same environment, but the impact of these results is only in one direction; i.e. the results of only one model are affected here. For example, an egress model may rely on the environmental conditions predicted by a fire model to predict smoke exposure to building occupants. Based on the smoke conditions predicted by the fire model, occupants within the egress model may reroute themselves to avoid the smoke or may become incapacitated resulting in a physical flow obstruction.

Two-way coupling is as above, but now the data affects the outcome within both (or multiple) models. For example, an egress model uses the environmental condition predicted by a fire model to influence evacuee movement and the fire model relies on the egress model to determine the architectural state (i.e. door positions) that in turn influences the movement of the fire and/or smoke. Currently, two-way coupling is not available outside of research products. The following questions might be asked of candidate models:

- *What fire information, if any, can be provided to the model such that it influences the outcome?*
- *How is this data provided?*
- *Do the generated results exist within the same environment allow effects to be explicitly represented?*
- *Am I able to determine when tenability might be an issue and make comparison with the evacuation results?*

The data entry required for some computer simulation models (especially those that represent individual evacuees, include behaviors, etc.) is more onerous than for those that only represent direct evacuee movement (including hand-based/manual engineering models). The user must therefore be cognizant of the resources and data available when considering the type of analysis that can be performed. The user should also be aware of the expertise required to configure the models available. A more refined and comprehensive model does not necessarily indicate a greater degree of effort to configure the model for a

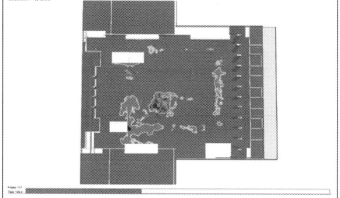 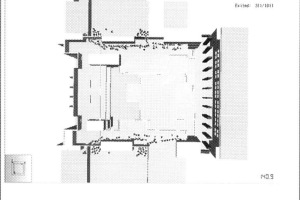

Fig. 9.2 Manual Overlay Analysis—Fire Model with points of visibility failure with Egress Model

Fig. 9.3 Automatic Overlay Analysis—Fire Model with points of visibility failure with Egress Model

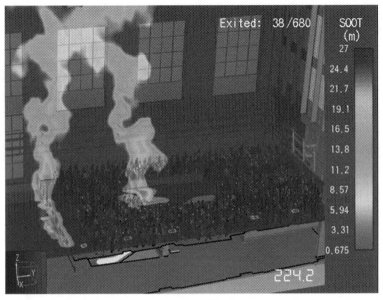

scenario. This will rely on the expertise available for the model and also the user interface. For instance, a simple text-based computer tool may have a command line interface, while the engineering calculations may require the manual manipulation of numerous equations in order to model even simple environments – possibly increasing the chances of error.

9.5 Model Output

The last element represented in Fig. 9.1 relates to the output that can be produced by the model. The model capabilities – whether it is computational or hydraulic in nature – can then be compared against the project requirements to determine whether the model output is sufficient to meet the needs of the project (in terms of analytical insights and output content) and the needs of the client/AHJ (in terms of the content, credibility and format of the output produced). This prompts a series of questions:

- What types of output can the model produced? For instance, numerical, tabular, still 2D images, 2D animations, 3D animations, VR output, etc.?
- What formats are used? For instance, does the output require dedicated tools to view it?
- How long does it take to generate the output? How long does the model take to run and is this sensitive to the size of the area/population being modelled and the type of output required?
- What level of output is produced? For instance, can output be produced on individual evacuees, groups, sub-populations or the entire populations? Can output be produced at specific locations, for component performance, sections/zones, floors or the entire building?
- Does the output reflect the summary conditions at the end of the scenario, or reflect timelines, specific conditions during the simulation process? (Refer to Chap. 10 for a more detailed analysis of the output types and its insights into model testing).
- Does the output capture key insights into the conditions produced during the simulation? For instance, engineering hand calculations provide high-level approximations at the level of the population within a space, while computer tools that represent individual agents can provide insights into individual occupant, their responses to a specific environment and the local conditions produced. Depending on the application, this level of detail is not always needed. The question then becomes, how significant an impact does this model simplification have on the credibility of the output produced?
- Example insights of particular interest might include:
  - *Merging flows (stair/horizontal):* For instance, the merging process at the floor-stair interface is especially important when specific floor clearing times are desired for a multi-floor building evacuation. Data on this dynamic is limited; however, studies by Galea [220] and Boyce [221] suggests the floor-stair merging process is sensitive to the configuration of the floor with respect to the incoming stair; specifically, if the floor is connected adjacent or opposite to the incoming stair.

- *Bi-directional flows* – consideration of the interaction with other evacuees, emergency services or support staff that may act as a contraflow to exiting occupants. Bi-directional flows commonly preclude the use of simple engineering flow calculations due to the lack of consistent empirical data required to develop a correlation.
- *High Levels of Congestion* – people-to-people interactions consider the physical characteristics of individuals within the occupant group indicating population densities at specific times and locations to judge comfort and safety levels. Hand calculations assume a homogeneous population within a space; therefore, local conditions within that space and their impact on an individual cannot easily be ascertained.
- Local navigation and evacuee path adoption.
- The performance of sub-populations with ranging movement and response capabilities and their impact on those around them; e.g. those using movement aids, with health issues/low fitness, with temporary conditions (e.g. broken limbs, pregnant, sick, etc.), long-term impaired, and/or older adults.
- The existence and maintenance of social groups during an evacuation and their impact on evacuee spacing, adopted travel speeds, route use, etc.
- The availability and impact of information on evacuee performance (e.g. notification systems, communication systems, guidance systems, non-emergency information, etc.).
- Output on the core evacuee performance components to determine their impact on an individual's/sub-population's performance; e.g. time to respond, the distances travelled, the congestion experienced, the safe arrival of population on an individual basis, the numbers using specific routes, exits, etc.

Ultimately, the engineer must decide upon the appropriate assessment method to be employed. This decision should be justified and documented for review. The strengths and limitations of the selected approach should be outlined along with any mitigating actions taken to compensate for them. The gap between the expected conditions and factors and those represented by the model should be clearly stated allowing a third-party to determine the credibility of the assumptions made and the results produced.

9.6 Characterization of Current Computer Based Evacuation Models

By identifying and understanding the key capabilities and characteristics of the computer simulation models, the user will be able to make a more informed choice between them. This analysis addresses issues of model refinement (i.e. the level at which the model operates), model source (i.e. the origins of the model), model distribution (i.e. how it might be accessed), and output type (i.e. what results can be produced) as identified in Fig. 9.1). In effect, this analysis helps the user answer the following question: *Does the model have the required attributes to credibly assess the scenarios of interest and to satisfy my project requirements?* The reader should refer to Chap. 8 for a detailed description of the hand-based (or manual) engineering approach.

Kuligowski [2, 222] outlined 26 computer-based evacuation models for buildings. She employed numerous categories to differentiate between the models. Of particular interest is whether they are movement models, partial behavior models, or behavior models. Movement models concentrate on the simulation of occupant movement and do not represent behaviors. Partial behavioral models primarily calculate occupant movement but also simulate some other evacuee behavior in an implicit manner. For instance, the implicit representation of pre-evacuation delays using time distributions, overtaking behaviors, and the introduction of smoke and its effects on the occupant movement speed. Behavioral models incorporate occupants performing a wider range of actions in addition to movement toward a specified goal (e.g. exit). These models may also incorporate elements of evacuee decision making and/or actions that are performed due to local conditions. In behavioral models, the behaviors simulated influence the evacuation performance of the individual or the population as a whole.

By definition, the information originally provided by Kuligowski [223] and later in the *SFPE Handbook of Fire Protection Engineering* [2] will become out of date within a short period of time, given the nature of model development. However, the categories used to differentiate between the models are likely to be a factor in model selection in the foreseeable future. The categories employed are described below and are employed in Table 9.1.

Table 9.1 Table of Computer Based Evacuation Models

| Model | Background of model | | | Model characteristics | | | | | |
	Developer/Institution	Validation	Availability	Modeling method	Refinement of population	Refinement of structure	Refinement of behavior	Output
EVACNET4	Kisko, Francis, and Nobel/ Univ. of FL, U.S.	FD	Y	M-O	Ma	C	N/A	T
WAYOUT	Shestopal/Fire Modelling & Computing, AU	FD	Y	M	Ma	C	N/A	V
STEPS	Mott MacDonald, U.K.	C, FD, PE	Y	M/PB	Mi	F	D	V
PedGo	TraffGo, Germany	FD, PE, OM, 3P	Y	PB/B	Mi	F	S	V
PEDROUTE	Halcrow Fox Associates, U.K.	N	Y/N3	PB	Ma	C	D	V
Simulex	Thompson/IES, U.K.	FD, PE, OM, 3P	Y	PB	Mi	Co	D	V
GridFlow	Purser and Bensilum/BRE, U.K.	FD, PE	Y	PB	Mi	Co	D	V
ASERI	Schneider/I.S.T. GmbH, Germany	FD, PE	Y	B-RA	Mi	Co	S	V
BldEXODUS	Galea and FSEG/University of Greenwich, U.K.	FD, PE, OM, 3P	Y	B	Mi	F	S	V
Legion	Legion International, Ltd., U.K.	C, FD, PE, 3P	Y	B	Mi	Co	S	V
FDS + Evac	VTT, NIST, Helsinki Univ of Tech	FD, PE, OM	Y	PB	Mi	Co	S	V
Pathfinder	Thunderhead Engineering	C, FD, PE, OM	Y	PB	Mi	Co	D	V
SimWalk	Savannah Simulations AG	FD, PE, 3P	Y	PB	Mi	Co	S	V
PEDFLOW	Edinburgh Napier University, Transport Research Institute	PE	Y	B	Mi	Co	S	V
SpaceSensor	Sun/de Vries	FD, OM	Y	B	Mi	Co	S	V
EPT	Regal Decision Systems, Inc.	FD	Y, N1	B	Mi	C, F, Co	AI	V
MassMotion	Arup	C, FD, PE, OM	Y, N1	B	Mi	Co	AI, S	V
Myriad II	Keith Still	PE, 3P	Y, N1	B	Mi	C, F, Co	AI	V
ALLSAFE	InterConsult Group ASA, Norway	OM	N1	PB	Ma	C	D	V
CRISP	Fraser-Mitchell/BRE, U.K.	FD	N1	B-RA	Mi	F	S	V
EGRESS 2002	Ketchell/AEA Technology, U.K.	FD	N1	B	Mi	F	S	V
SGEM	Lo/University in Hong Kong	FD, OM	N1	PB	Mi	Co	D	V
EXIT89	Fahy/NFPA, U.S.	FD, OM	N2	PB	Mi	C	D	T
MASSEgress	Stanford University (Civil and Env Engineering)	PE, OM	N2	B	Mi	Co	S	V
EvacuatioNZ	Spearpoint/ University of Canterbury, NZ	FD, PE, OM	N2	B	Mi	C	S	V

The various category types are:

- *Validation: C = Validation against codes; FD = Validation against fire drills or other people movement experiments/ trials; PE = Validation against literature on past experiments (flow rates, etc.); OM = Validation against other models; 3P = Third-party validation; N = No validation work could be found on the model*
- *Availability to the Public: Y = The model is available to the public for free or a fee; N1 = The company uses the model for the client on a consultancy basis; N2 = The model has not yet been released; N3 = The model is no longer in use; U = Unknown*
- *Modeling Method: M = Movement model; M-O = Movement/optimization models; PB = Partial behavioral model; B = Behavioral model; B-RA = Behavioral model with risk assessment capabilities; B-AI = Behavioral model with artificial intelligence capabilities;*

9.6 Characterization of Current Computer Based Evacuation Models

- *Refinement of the Population: Ma = Macroscopic; Mi = Microscopic*
- *Refinement of Structure: C = Coarse network; F = Fine network; Co = Continuous*
- *Refinement of the Behavior: D = Deterministic; S = Stochastic; AI – Artificial Intelligence; N/A = Information not available*
- *Output: T = Textual output; V = Visual output.*

This table is provided only as a baseline model assessment – as an indication of the types of models available and the insights that the categories can provide to the reader into the model attributes. *The material included in the table should only be considered indicative and certainly requires further confirmation.* The model user is therefore encouraged to gather further information for any candidate model to ensure that the model has not changed and that it is still available.

This section has provided a set of questions that the prospective user should examine when considering candidate egress models. These related to project requirements and constraints, model attributes, model scenarios to be investigated, and the output required. It is suggested that a potential model user should review these questions when considering the selection of an egress model. The questions may not always be of equivalent importance, depending on the application type and project demands. However, establishing which are important and then answering them will enable comparison between candidate models and enable a more evidence-based selection process.

Egress Model Testing

10

10.1 Introduction

This chapter is intended to provide guidance for **egress model users** regarding:

- The selection of an appropriate egress model
- The configuration of the selected egress model for the scenarios of interest
- The testing of the configured egress model to assess whether it is fit for purpose
- The reporting of test results.

It is hoped that this guidance will enable a more informed approach to selecting and applying a model, while providing a third party with more confidence in the results produced.

Computational egress models are often applied to establish the time for an evacuating population to reach safety as part of a performance-based design. Egress models are also used to extend research envelopes, examine the evacuation dynamics of different scenarios and examine the effectiveness of procedural/structural developments [217]. These models form an increasingly important aspect of engineering practice and therefore warrant scrutiny of their performance and the assumptions on which they are based. This section presents guidance for the user to test the performance of these models and examine the underlying assumptions.

Prior to employing a model, the model user needs to determine if the model is capable of generating *useful*, *appropriate* and *credible* results. The formal process by which this is demonstrated is Verification and Validation (V&V). A number of definitions of verification and validation are available. As an indication of the definitions employed, the American Institute of Aeronautics and Astronautics (AIAA) [224] states:

- **Verification**: *The process of determining that a model implementation accurately represents the developer's conceptual description of the model and the solution to the model.*
- **Validation**: *The process of determining the degree to which a model is an accurate representation of the real world from the perspective of the intended uses of the model.*

These two definitions are adopted here. In their most basic form, *validation determines whether the right model is being used* for the application in question, while *verification examines whether this model has been implemented as intended and properly portrays the intended phenomena*. Validation and verification are not equivalent or interchangeable and therefore both forms of testing are necessary. Given that egress models can often be reconfigured by the user—modifying the model assumptions[1]—it is important that the user can both confirm that the model is functioning as planned and that the plan reflects reality to a reasonable degree.

---

[1] And therefore the model itself.

© Society of Fire Protection Engineers 2019

SFPE Society of Fire Protection Engineers, *SFPE Guide to Human Behavior in Fire*, https://doi.org/10.1007/978-3-319-94697-9_10

97

The use of model results requires qualification and justification to ensure that the model limitations are understood and taken into account. Ideally, all models would be thoroughly tested for all potential scenarios and parameter permutations. This is practically impossible given the array of model parameters and scenarios typically available. This implies that there will inevitably be gaps in model testing. This becomes even more likely given the increasing range of scenarios to which a model might be applied (see Chap. 6). Model developers typically have access to more model components than a user, given their more intimate access and understanding of the model and its internal logic. However, even with this increased access and their interest in ensuring model performance, a developer will still not be able to address all eventualities. The model user should then perform tests to:

- Confirm the reliability of previous relevant tests performed by other parties
- Confirm their own understanding of the model and expertise in applying the model
- Ensure that the model has been sufficiently tested for their current scenarios of interest.

Model use then needs to be clearly understood and justified given the application at hand. This is just as true for egress models as it is for any other model.

This guidance is intended for model users. Here, a simple difference is assumed between a model user and a model developer. A model developer is assumed to have access to all aspects of the model (including the internal workings of the model), and have a high functional knowledge of the model's structure. A model user is assumed to have limited access to the internal workings of the model (and can then only affect the results by configuring the model via the interface provided), and whose understanding of the model may vary given their experience and training.[2] This discussion is intended to aid the user in judging existing model testing and to conduct their own testing in order to increase their confidence in the model's performance and the credibility of the results produced; i.e. to ensure the *net level of testing* performed and subsequent confidence in its application to the scenario(s) at hand. It should be noted that testing conducted by the model user is an important safeguard to enhance credibility; however, it does not replace developer testing. The user should be highly sceptical of a model that has no publicly documented testing provided by the developer [225–228].

10.1.1 Relevant Work in Fire

The fields of computer science and mathematical modeling have produced a number of documents on model testing [229–241]. These provide a detailed, although not always consistent, description of the approaches available. In addition, several authors have produced testing guidance specific to fire models (e.g. Dey [242], Mok [243], and Dalmarnock [244] etc. [245–247]). These authors provide useful insights into the testing approaches adopted in fire safety. There has also been some guidance developed specifically for testing egress models. These include the contributions of Ronchi et al. [248] Lord et al. [225] Galea et al. [249] and the regulatory guidance provided by IMO for egress models used in a maritime environment [250]. These contributions informed the development of this section and are referred to throughout. The reader is encouraged to examine the work of Lord et al. [225] and Galea et al. [249] to examine their original approaches.

This chapter first outlines the process of model application. It then presents tests that can be performed before the model is executed (informing model selection and configuration) and after the model is executed (model verification and validation). Finally, this section identifies several techniques that can be employed to assess model performance. This guidance is intended to suggest alternatives for consideration by the user. In addition, a recommended *minimum net level of testing* is suggested. The user's final decision, as to which tests are performed and how they will be performed, will be dependent on the information and resources available. However, it is critical that the model user understands the testing opportunities, the techniques available and can justify the decisions taken.

---

[1] And therefore the model itself.

[2] In reality, every time a user configures a model s/he is effectively developing a new model; however, the distinction between model user and developer is maintained here for simplicity.

10.2 Testing the Process

Modeling is a process that typically involves actions performed by several parties (e.g. data collectors, subject matter experts, model developers, users and third-party reviewers), numerous resources (e.g. the basic model, data, scenario definition, etc.) and passes through a number of stages (e.g. data collection, theory development, model development, data selection, model configuration, model testing and calibration, model application, and interpretation of the results produced). The focus here is on the stages available to the user for them to determine the appropriateness of the model to the task at hand:

<u>Pre-Model Execution</u>

Step A. Model Selection—*Deciding on the model to be used from the material available.*
Step B. Model Configuration—*Modifying the model (settings) to better represent the scenario of interest.*

<u>Post-Model Execution</u>

Step C. Model Verification—*Comparing model performance and model expectation.*
Step D. Model Validation and Calibration—*Comparing model results and real-world conditions to ensure that they are sufficiently similar, and then refining model settings such that results better approximate real-world conditions.*

Each of these stages requires the user to understand and assess the model, the model's performance and its potential use; i.e. to establish the model's ability to be applied to the scenario at hand and generate useful output. These stages will now be briefly addressed in turn, although in reality this is not a linear process and may involve numerous iterations.

This section identifies egress model tests that might be performed, the techniques available to examine and/or compare the results produced, and the test information that should be reported to interested parties. This is presented in order to provide the user with testing options and some basis for selecting between these options. A **minimum combined level of model testing** is identified. This minimum combined level may be created by the model user or may have already been produced by a third party. This net level of reported testing is critical in demonstrating the credibility of the model being applied to the scenario of interest, and the user's understanding of this performance. An outline of the minimum testing report is shown below. Each of the stages is described in more detail in the following section.

> **Minimum Combined Level of Model Testing**
> **Model Selection—*Justify the selection and use of the model.***
>
> - *What documentation is publicly available describing the model's assumptions and functionality?*
> - *What documentation is publicly available describing previous model testing?*
> - *What aspects of the model are driven entirely by user actions as opposed to model forecasts?*
> - *What techniques are employed by the model to represent key modeling components?*
> - *What evacuee behaviors are represented by the model?*
> - *What output is produced by the model?*
>
> **Model Configuration—*Identify the factors being examined, scenarios formed and the model changes required to reflect these factors.***
>
> - *What real-world factors might influence evacuee performance?*
> - *What is the expected impact of these factors on evacuee performance?*
> - *What model changes have been made to represent the impact on evacuee performance?*
>
> **Model Verification—*Confirm that the model changes have been made and the model performs as expected given the new settings.***
>
> - *Do assigned pre-evacuation times lead to agents responding at the prescribed time?*
> - *Do assigned travel speeds lead to agents moving at the desired speed given the various terrain types faced?*

- *Do the measured or imposed peak flow rates for key geometric components limit the achieved flow rates through the egress component as specified for the given scenario?*
- *Do agents use assigned routes? Do agents avoid routes that are deemed unavailable?*

Model Validation—*Confirm that the model's predictions for the scenarios of interest are not invalid.*

- *What is the source of the benchmark data employed?*
- *What is the event scenario from which this data was collected?*
- *What is the format of the benchmark data?*
- *What changes to model inputs have been made and how do they reflect the benchmark scenario?*
- *What measures are being compared in the test?*
- *What comparative tests are being applied and what assumptions do they make/require?*
- *What acceptance criteria are adopted for these measures and why?*
- *What are the resultant differences between the simulated and benchmark data (paths adopted, exit use, exit arrivals, overall evacuation times)?*
- *What are the implications of the results given the acceptance criteria identified?*

10.3 Pre-Model Execution

10.3.1 Step A – Model Selection

For the model user, the selection of the model is a key influence over the modeling process and represents the initial test of the model's suitability. In reality, model users will not all have the same degree of influence over model selection and the scenarios to investigate; e.g. their organization might already have a model license, or in-house resources, expertise, etc. Irrespective of the influence over model selection, the user should still justify the application of the model to the scenario of interest. Depending on the circumstances, the user may perform tests to *select* a model or to *justify* the use of a given model. This allows a third party to determine the rationale behind the application of this model to the project at hand. The user should then ask a number of questions[3] to test the model's appropriateness:

[MS1]. Does the model have sufficient technical documentation for the user to understand the model assumptions (e.g. theories, data-sets used, etc.)? For instance, can the user assess that the conceptual model employed is sufficiently representative of the scenario being represented?

[MS2]. What was the intended application for the model? How is this reflected in the default settings of the model and what modifications to the default settings/parameters need to be made to configure the model for the current scenarios?

[MS3]. Does the model have sufficient documentation for the user to assess previous model testing? Does the user have sufficient information to judge the current credibility of the model and the testing activities that are still outstanding? Is it considered an acceptable tool by the approving authority having jurisdiction for the project?

[MS4]. What is the user-relationship with the model and sub-components?
- User-driven—can the user dictate how evacuees respond and perform (how much control does the user require and how will they justify the assumptions made)?
- Predicted—does the model determine how evacuees respond and perform?

[MS5]. Does the user have sufficient expertise to employ the model? If not, is relevant training available? To what extent is technical support provided by the model developers to operate the model? Will the user be able to use the model in a demonstrably credible manner?

[MS6]. Can the model actually be employed to the application at hand?
- Can it cope with the scale of the project? For instance, the number of people, the size of the geometry, etc.

---

[3] Or establish that these questions have already been asked.

10.3 Pre-Model Execution

- Can it be configured to reflect the target scenarios? For instance, the procedure employed, the nature of the incident, etc.
- Can it be configured and executed in the time available? Does it require specialist resources to be used? Does it require large numbers of simulations given the nature of the model (stochastic, deterministic, etc.)?

[MS7]. How is the model distributed and how much does it cost? For instance, is it free, open source, in-house, licensed, etc.? Will this distribution approach allow it to be used in the current project?

[MS8]. Are appropriate techniques employed within the model to represent core elements of the evacuation process in sufficient detail?

- How is the population represented? For instance, as a flow, as uniform individuals, as individuals defined by a set of attributes, etc.
- How is the building represented? At what level of refinement does the model represent the space around which the population can move (e.g. a continuous space, tiles, between rooms, etc.)?
- How are the environmental conditions represented?
  - ◦ Does it represent the deteriorating conditions?
  Using empirical data?
  Using user-derived functions?
  Using input from a zonal model?
  Using input from a CFD model?
  - ◦ What impact do these conditions have upon performance?
  The well-being of the population,
  The routes available,
  The decision-making of the population?
  - ◦ Given these considerations, is the user able to represent the scenarios of interest in sufficient detail?

[MS9]. How is the emergency procedure and evacuee decision-making process represented? At a minimum, how does the model address the following key performance elements (see Chap. 6):
- Pre-evacuation delays
- Travel speeds (and possible delays in movement towards safety)
- Flow constraints
- Route availability and use
- Could a suitably configured version of the model reflect the scenarios of interest?

[MS10]. What output does the model provide?
- Does it address the key subject matter? For instance, evacuation times, distances travelled, congestion experienced, routes used, hazard exposure levels, etc.
- Is it in the desired format? For instance, numerical, descriptive, graphical snapshots, 2D/3D animations, immersive, etc.
- Is it at the required level of refinement? For instance, at the agent level, component level, floor level, building level, scenario level, etc.
- Is the output sufficient for the current application?

The user should be able to satisfactorily answer these questions prior to making (or justifying) their model selection. The answer to these questions should be reported in conjunction with the results produced by the model—as justification for the model's selection and use.

Minimum Set of Model Selection Questions

*It is recognised that the user may not have the means to address and report all of these questions. It is therefore suggested that **as a minimum** the following questions are addressed and reported.*

- *What documentation is publicly available describing the model's assumptions and functionality?*
- *What documentation is publicly available describing previous model testing?*

> - *What aspects of the model are driven entirely by user actions as opposed to model forecasts?*
> - *Does the model have sufficient documentation for the user to assess the model testing previously performed?*
> - *What techniques are employed by the model to represent key modelling components?*
> - *What evacuee behaviours are represented by the model?*
> - *What output is produced by the model?*

OUTCOME: The model user has selected an appropriate model for the project.

10.3.2 Step B – Model Configuration

Given that the user has selected a model, they must then configure the model for the scenario(s) of interest. This requires the user to examine the real-world conditions associated with the scenarios and represent the initial conditions (i.e. the structure, the population, the procedure, the environmental conditions, etc., described in MC1 below) and the evacuee response to these initial conditions (described in MC2-4 below). We focus on the latter here. This will require users to update the conceptual model of evacuee response (where the conceptual model is the combined set of evacuee behavioral and performance assumptions made that are built into the model), by configuring the model settings to more credibly reflect the scenario (s) of interest. That is the user attempts to at least ensure that the key elements in point [MS9] above are more credibly represented. This may require the user to provide data and/or update model settings to manipulate model assumptions (the reader is referred to Chap. 6 for guidance on the development of egress scenarios). This will involve the following steps:

- *[MC1] Identify Scenario Factors: What real-world scenarios are being examined and what scenario factors (e.g., voice notification system is in place, elderly population are present) are employed to produce the scenarios of interest* [166]?
- *[MC2] Identify Expected Impact: What is the expected impact of these factors on evacuee performance* [101]? Many factors can influence evacuee performance. These factors can be grouped according to four elements typically represented in engineering analysis (pre-evacuation delays, travel speeds, flow constraints, and route availability/use, see Chap. 6), which can be addressed in the majority of egress models. Although a crude simplification, these four performance elements do enable the key physical aspects of an evacuation to be represented and should, at a minimum, be considered.
- *[MC3] Identify Model Default Settings: What default model settings are made that are relevant to the scenarios of interest?*
- *[MC4] Suggest Model Changes: What information is available to support model configuration of the four elements of evacuee performance?*
- *[MC5] Actual Model Changes: How has the model been configured to better reflect scenario conditions (in accordance with MC4; e.g. GUI parameter settings, model input file, etc.)? What changes were actually produced through the suggested changes to the model?*

In following these steps, the user is demonstrating an understanding of the discrepancy between a model's default assumptions (the default conceptual model of evacuee performance) and the expected evacuee behavior during the scenario of interest, and then justifying their actions to bridge this gap. An example of how this process might occur is shown in Table 10.1.

This represents the transition from the real-world conditions to the simulated scenarios of interest. It allows third party viewers to examine the steps taken in this process, the assumptions made and the eventual similarity between the real-world and simulated conditions. The user should **verify** that the new model configuration has been implemented as planned (i.e. compare MC5 with MC4). The user should then **validate** the newly configured conceptual model to confirm it better represents their understanding of the scenario factors (i.e. compare MC4 with MC2); i.e. that the configured model more effectively represents expected evacuee response than the default settings.

10.4 Post-Model Execution

Table 10.1 Reporting of Model Configuration

MC1 Scenario Factor →	MC2 Expected Impact→	MC3 Model Default → Setting	MC4 Suggested Model → Changes	MC5 Actual Model Changes
Voice Notification System in Place	Reduction in Pre-Evacuation Time. Expected pre-evacuation time between x-y sec	0 s	Represent Lognormal Distribution of pre-evacuation times between x-y secs	Generate a screen-grab of population generation given attribute settings.
Elderly Population Present	Increase in pre-evacuation times, reduction in travel speeds. Expected pre-evacuation time between x-y sec Expected travel speed between a-b m/s	0 s 1–1.5 m/s	Represent Lognormal distribution of pre-evacuation times between x-y secs Normal distribution of speeds between a-b m/s for sub-population.	Generate a screen-grab of the population generation and distribution produced by the model given the attribute settings provided.

> **Minimum Level of Model Configuration.**
>
> *It is recognised that the user may not have the means to address and report all of these questions. It is therefore suggested that **as a minimum** the following questions are addressed and reported:*
>
> - *What real-world factors might influence evacuee performance?*
> - *What is the expected impact of these factors on evacuee performance?*
> - *What model changes have been made to represent the impact on evacuee performance?*

OUTCOME: The model user can demonstrate that the new model configuration has enhanced the model's representation of the scenario of interest.

10.4 Post-Model Execution

The following tests require the model to be executed; i.e. run and applied to a test case. The configured model is applied and the results examined to determine whether they are as specified and/or are sufficiently representative of real world conditions, depending on the nature of the tests.

Egress models operate on a number of levels and can produce results that reflect these levels. Depending on the nature of the model, it might operate on the:

- *Individual level*: simulated evacuees[4] perform actions. For instance, what route did they adopt, how quickly were they moving, how far did they travel, what time did they reach an exit, etc.?
- *Aggregate level*: simulated evacuees interact with each other or with simulated objects that then influences the local conditions in a measurable way during the simulation. For instance, how long did agents queue at a particular exit, where did flows merge, what levels of congestion were observed, etc.?
- *Scenario level*: where results summarize the conditions at the end of the simulation. This can apply to a population, a component, an area, the entire building, etc. For instance, how long did it take to evacuate the building or floor, which exit had the greatest use, how far did a group of agents with mobility impairments have to travel, etc.?

Access to these three levels of results allows the user to understand the underlying dynamics during the simulation and the consequences of these dynamics across the simulation. This information allows the user to explore the relationship between the conditions experienced by the simulated agents at different times and/or locations and the overall outcome of the scenario.

---

[4] Typically called agents.

This will help the user determine whether represented underlying dynamics produce credible outcomes; i.e. allow comparison between relationships between measures as well as the examination of measures in isolation. This allows the identification of:

- influential factors
- the level at which these factors operate
- the nature and extent of their impact
- the relationship between factors

This information also allows the user to make comparisons both within a scenario (across different locations, different times, events, etc.) and between different scenarios.

Several examples of the information that an egress model can produce are shown in Table 10.2 for each type of level. Although not exhaustive, these examples are indicative of the output that current egress models can produce.

It is not suggested that all of these different results can and should be reported for every scenario examined. Indeed, some models and some scenarios may not generate all of these results. However, the model user should be aware that results are produced at these different levels and that they provide different insights; therefore, excluding levels might preclude insights into model performance. This weakens the insights provided and the claims made of the testing procedure.

In the next two sections, post-execution verification, validation, calibration and sensitivity analysis are discussed. These tests involve the comparison of the measures shown in Table 10.2, and can therefore establish the qualitative and quantitative similarity between the simulated conditions, model specification and/or expected real-world conditions.

It is not suggested that the user should perform all of the following verification and validation tests every time that they employ an egress model. That would be too onerous and may produce a great deal of redundant information. For instance, model tests may already exist that relate to the scenario of interest. The decision to perform a specific test should be informed by the following questions:

- *Have tests been performed before?*
- *How well were they documented?*
- *Who performed them and what was the expertise of those involved?*
- *How relevant was the scenario tested to the current scenario?*
- *What measures were examined during the test?*
- *What conclusions were drawn?*

These questions allow the user to determine whether existing tests provide sufficient confidence in the model's suitability for the current application. However, reliance on existing testing removes an opportunity for the user to demonstrate their own expertise in model configuration and model use. This may not always be necessary, but can be a useful side effect of documented, detailed, and expert model testing. This examination helps the user determine the existing and required testing to ensure that the net level of testing produces sufficient confidence in the model's performance.

Although tests might not be performed every time a model is to be applied as part of a project, it is good practice to state why the model user is confident in the application of the model to the scenarios of interest and cite evidence for this confidence; i.e. either their own current tests or credible previous tests.

Model verification can be conducted at the individual, aggregate and scenario levels. Doing so provides different insights—insights that enable the user to assess how the simulated conditions progress over time and across the simulated space.

10.4.1 Step C – Model Verification

If the previous steps have been followed (i.e. Steps [A] and [B]), then the user-configured conceptual model (i.e. behavioral and performance assumptions) should now better represent the scenarios of interest. The performance of this model needs to be verified to establish that the basic functionality performs as expected (i.e. whether the configured model represents the user's intentions).

Verification tests allow comparison between model performance and model specification. The user might produce simple test cases to explore this relationship. This has been highlighted in detail by Ronchi et al. [248] and previously in IMO [251] documentation. These tests can be aimed at the individual level (e.g. agent travel speed), aggregate level (e.g. the congestion

10.4 Post-Model Execution

Table 10.2 Example Measures for Individual, Aggregate and Scenario

#	Individual level results (I)	Aggregate level results (A)	Scenario level results (S)
1	[I1] Evacuation (or action) Time	[A1] Flow Conditions (Existence/Extent/Location)	[S1] Summary Evacuation Times (Floor/Space/Building/Sub-Population)
2	[I2] Distance Travelled	[A2] Flow Patterns/Agent Grouping (Nature/Location)	[S2] Injuries/Fatalities (Number/ Location/Time)
3	[I3] Time Spent in Congestion	[A3] Population Distribution (Location/Size/Densities)	[S3] Summary Flow Rates (Component(s))
4	[I4] Pre-Evacuation Time		[S4] Summary Congestion Levels Experienced
5	[I5] Path(s) Adopted		[S5] Summary Distances Travelled
6	[I6] Component(s) Used (Egress Component/Route/Exit)		[S6] Summary Components Use
7	[I7] Physiological Status (FED)		[S7] Summary Pre-Evacuation Times
8	[I8] Achieved Travel Speed(s)		

that develops at a bottleneck, given prescribed flow rates), and even the scenario level (e.g. the time it takes an agent population to arrive at an exit given that the speed, flow constraints and distances traversed are all known). This involves comparison against expectation, not against reality. The measures presented in Table 10.2 can then be used as part of the verification process. Test cases can be constructed that focus on the simulation of these measures; the results can then be compared with user specification. These tests typically require the user to constrain agent performance, such that only one or two variables depend on the scenario conditions, while the other variables are dictated by the user. A number of example model verification tests employing the measures identified in Table 10.2 are listed below (with the *MVf* prefix added to the original measure label to indicate model verification):

- *[MVf_I1] Evacuation (or action) Time/Location—Does an agent reach a specific location at the expected time given their assigned travel speed?*
- *[MVf_I2] Distance Travelled—Given a specified route, is the correct travel distance recorded for the agent?*
- *[MVf_I4] Pre-Evacuation Time—Does the agent respond at the specified time?*
- *[MVf_I5] Path Adopted—Is the path adopted credible (e.g. agent movement does not breach a wall or an obstacle) and as specified?*
- *[MVf_I6] Component (Exit) Used—Does the agent use the exit specified?*
- *[MVf_I7] Physiological Status (FED)—Does the FED model accurately record agent exposure to the environmental products given the environmental conditions faced?*
- *[MVf_I8] Attained Travel Speed—Does the agent travel speed change to reflect the current terrain/conditions/activity as expected?*
- *[MVf_A1] Existence/Extent/Location of Congestion—Does a queue form when a component is overloaded? For instance, when 50 agents arrive at a 1 m exit with flow capped to 1p/s?*
- *[MVf_A3] Population Densities—Are population densities accurately recorded and are they reasonable? For instance, do population densities fall below the model maximum of $4p/m^2$?*
- *[MVf_S2] Number/ Location/Time of Injuries/Fatalities—Do the number of fatalities and injuries reflect expectation given the conditions represented; i.e. does the timing of the events reflect the implemented FED equations?*
- *[MVf_S3] Summary Flow Rates—Do recorded component flow rates meet expectation?*
- *[MVf_S7] Pre-Evacuation Distribution—Do the agent pre-evacuation times lie within the specified range and follow the specified distribution?*

For simplicity, only examples involving the core performance elements are shown here. In practice, the user should verify the aspects of the model that have been modified to reflect the scenarios of interest, along with those aspects expected to influence the simulated outcome.

> **Minimum Level of Model Verification**
> The user may not have the means to address and report all of these verification activities. It is therefore suggested that as a minimum the questions are answered and reported.
>
> - *Do assigned pre-evacuation times (or distributions) lead to agents responding at the prescribed time?*
> - *Do agents use assigned routes? Do agents avoid routes that are deemed unavailable?*
> - *Do assigned travel speeds lead to agents moving at the desired speed given the various terrain types faced (e.g. corridor, stairs, etc.).*
> - *Do the measured or imposed peak flow rates for key geometric components limit the achieved flow rates through the egress component as specified for the given scenario?*
>   *These questions should be answered irrespective of whether the user is reporting tests conducted by them or tests previously conducted by third parties.*

OUTCOME: The user verifies that the model is performing as specified.

10.4.2 Step D – Model Validation and Calibration

The user must determine that the model's predictions are credible (i.e., representative of real-world conditions). At this stage, **model validation** requires comparison between simulated results and real-world conditions. Model execution is required to achieve this. Operational validation [229–236–241, 252, 253] establishes whether the model output produced after execution is representative of relevant real-world outcomes and is formed from:

1. *Outcome Validity*—representative outcomes are produced given specified initial conditions (Individual - > Scenario Level)
2. *Progress Validity*—representative conditions evolve during the simulation given specified initial conditions (Individual - > Individual/Aggregate Levels)
3. *Event Validity*—the number, type and location of events emerge during the simulation that are representative of real-world conditions given specified initial conditions (Individual - > Aggregate Level).

Tests are outlined in the following sections to examine these different types of operational validity.

As with the verification tests, comparison is required at the individual, aggregate and scenario levels (see Table 10.2). Each of these levels provides different insights into the simulated conditions and allows different comparisons to be made with real-world conditions. Many of the techniques presented below can be applied to outputs from these levels allowing examination of each level of output and the relationship between them.

It is assumed that these tests are typically open (*a posteriori*); i.e. that the user is aware of the scenario conditions present during the benchmark case and the outcome prior to the execution of the model during the test. If this data is not available to the user (i.e. a blind or priori test) then the user's skill at identifying and representing real-world factors becomes an influential factor in the results produced. Although this is an important type of testing that provides an array of important information, it does introduce an additional variable into the output, somewhat complicating the analysis of the results produced (i.e. it tests the modeling process as much as the model capabilities). Therefore, the primary model user testing is assumed to be open, with blind testing assumed performed only when the model has first been subjected to open tests in order to determine the impact of the user's prior knowledge on the outcome [242].

It may be that the results of a particular test can be improved. This might be achieved by minor modifications to the model's settings, whilst still representing the same scenario of interest. The model can be **calibrated** in this way to further hone the operation of the model—to improve the validity of the model by making minor adjustments to the configured model, while maintaining the key model assumptions.

10.4.2.1 Benchmark Data
Prior to comparison, a benchmark data-set first needs to be selected. Simulated results may be compared against a number of different types of benchmark data that are assumed to approximate/reflect real-world conditions (e.g., a real incident, experimental trials, evacuation drills, engineering judgement (i.e. where the user defines expected performance according

10.4 Post-Model Execution

to their experience and expertise, rather than according to related numerical evidence), or from the simulated results produced by other models, ideally those that have been rigorously subjected to the validation process). The value of this comparison is highly dependent upon the perceived validity of the benchmark case employed. It is also dependent on the proximity between the simulated and underlying benchmark scenarios; there is little value in approximating the benchmark data unless the underlying scenarios are equivalent.

The source of the benchmark case will influence the perceived credibility of the data used and therefore of the comparison made. *Real incident data* are typically assumed to be the most credible sources for comparison. However, the data-sets are likely to be partial (given the lack of data collection preparations), more qualitative (given the reliance upon reported conditions), and unlikely to be directly relevant to the scenario of interest. Examples of real-world incident data-sets include: [51, 80, 82, 254–258]. The user should examine the limitations of the data-sets available in each case, along with the contextual information available that would allow the user to establish the similarity between the data-set scenario and scenario of interest. Gaps in the data-set would limit the comparisons made or would need to be filled with other data/user assumptions.

An *egress exercise* (e.g. a routine evacuation drill) may potentially produce a more complete data-set than real incident data, given the opportunity for those collecting the data to thoroughly prepare and instrument the building/situations of interest. However, the exercise may not be sufficiently realistic given the lack of a triggering incident in the exercise (e.g. no smoke)[5] and the potential for those involved having prior knowledge that the exercise will take place. There are a number of example data-sets that relate to egress exercises [37, 59, 206, 259–262]. Data-sets are also available that describe the observation of *routine pedestrian movement* (e.g. pedestrian circulation around an observed space).

Experiments allow researchers to control conditions and focus on particular aspects of participant performance. These are normally set-pieces and so are either artificial in nature (i.e. with the participants aware that they are taking part in an experiment) or focus on controllable aspects of the evacuation process (e.g. travel speeds, pre-evacuation times, etc.). As such, it might inform aspects of a scenario of interest, but is unlikely to provide sufficient data to describe the whole process.

The *simulated output of other egress models* might allow detailed comparisons to be made (e.g. of the underlying conditions and the outcomes produced [229–241], assuming that the models produce comparable types of output and are configured to represent the same scenarios. The credibility of this comparison is highly dependent on the benchmark model, the scenario examined and, as in all cases described here, the relevance of this to the scenario of interest. Several examples of this type of comparison are available.

Where none of the previous options are available, the user might need to rely on their own analysis—where estimates are extracted from available theory and the data available to produce a composite data-set or expected outcome. For instance, the user determines that travel speeds should be reduced by X% across a novel floor covering based on data-sets reflecting similar, but not identical, surface conditions. The validity of this approach is largely dependent on the basis of these insights and the case made by the model user to justify the benchmark case developed—whether these insights came from subject matter experts, engineering guidance, standard practice, etc. Again, as in all of the approaches presented, it is critical that this process is as transparent and well-documented as possible. In some situations, the theoretical and empirical support for this analysis will be weak, requiring the user to provide an opinion on the expected performance (e.g. engineering judgement).

Current data-sets are typically quite limited—in their completeness, consistency, refinement, and description. This significantly hampers the testing process. The effort required on behalf of the user to gather and configure appropriate data should not be underestimated. In many instances, the user will not be able to use the original data-set directly, but will instead be required to modify/augment/compile the data in some form. The combination of different data-sets and the potential user actions that might be required are too large to provide comprehensive guidance here. However, one guiding principle is for the user to be as transparent as possible in their assumptions and their actions.

Minimum Level of Benchmark Data
What is the source of the benchmark data?
What is the event scenario from which this data was collected?
What is the format of the benchmark data?
What model changes have been made and how do they reflect the benchmark scenario?

---

[5] It should be noted in real-incidents some evacuees may not be exposed to fire effluent and therefore may be similarly reliant upon notification as are evacuees during an egress drill.

Table 10.3 Selection of the comparison techniques available

| Technique | Purpose | Test applications | Type of test | | Example test outcome[a] |
			Qualitative	Quantitative	Level
[CT1]	*Is the Behavioral logic employed based on credible assumptions?*	Pre-Execution[b]	Visual inspection of flow chart depicting agent decision-making process		Agent decision-making process
[CT2]	*Do the agent behaviours appear superficially credible given the conditions faced?*	Post-Execution	Visual inspection of narrative/ stills/animated playback		I5, A2-A3
[CT3]	*Do events occur when and where they are expected?*	Post-Execution	Visual inspection of event timeline		I1-8, A1-A3
[CT4]	*Do events occur as frequently as expected?*	Post-Execution	Visual Inspection of histogram/table	Statistical Comparison of Frequencies	S1-7
[CT5]	*Do conditions/events evolve as expected?*	Post-Execution	Visual inspection of data trace/time series curve	Functional analysis of result convergence	I1-I3, I5-I8, A1-A3
[CT6]	*Are relationships between variables as expected?*	Post-Execution	Visual inspection of Behavioral diagram/scatter-plot/trend-lines	Correlation analysis	Compare I2/I3/ I7 vs. I1 I8 v I3/A3
[CT7]	*Do simulated and real-world measured distributions overlap to an acceptable degree?*	Post-Execution	Visual Inspection of box-plot/tabular data	Confidence interval analysis	S1-S7
[CT8]	*Are simulated and real-world performance measures sufficiently similar? Assess similarity between two or more variables.*	Post-Execution	Subject matter expert test	Statistical Comparison of Performance Measures	S1-S7

[a]Refer back to Table 10.2.
[b]Although strictly conducted without model execution, it is typically conducted when a model is implemented and capable of execution.

10.4.2.2 Comparison Techniques.

A number of techniques are available that allow comparison between model results and benchmark data-sets (i.e. to enable the previously described validation and verification tests to be performed) [225, 229, 231–234, 236, 238–240, 248, 263–268]. A brief overview of these techniques is provided below in Table 10.3. In reality, the resources, (computational, human and project-based), and the information available will largely determine which of the approaches are adopted. Guidance is focused on which of the techniques is most suited for the tests outlined previously given the level of model assessment (see Table 10.3).

A number of approaches can be used to assess the model performance (see Table 10.3). These are identified as either being qualitative (i.e. judgment-based comparison between simulated and real-world conditions), or quantitative in nature (i.e. involve statistical or numerical comparison). Although the latter approaches may generally be considered more rigorous, they require more effort, will likely require more data and may also require a number of conditions to be met prior to their use (especially when involving statistical analyses). Given that model users have limited resources to conduct tests, it is unlikely that they will be able to employ all of the approaches outlined but should still justify the approaches adopted. Means to qualitatively assess results includes examination of the following (described in detail by Sargent [229, 231–234, 236, 238–240]):

- A **flow chart**—the logic of an agent's decision-making process. For instance, a graphic depicting agent behavior in response to smoke interaction is compared with a relevant case study to establish whether the simulated logic is credible.
- A descriptive/graphic **narrative**—a description and/or series of graphics that reflect the development of conditions over time. Allows comparison between predicted developments and actual conditions. For instance, is the shape of the crowd around an exit representative of photographs of congestion?

10.4 Post-Model Execution

- A **timeline** graphic—allows the model user to chart the occurrence, frequency and relative position of events on an axis (or schematic) and compare them with actual event timelines. For instance, does a simulated queue deplete at the expected time given real-world observations?
- A **histogram** can be used to display frequency and relative frequency distributions. For instance, does the number of exit uses represent expectation?
- A **behavior diagram/scatterplot** enables two variables to be plotted against each other for a scenario in order to identify the nature of the relationship between them. For instance, does the exposure to elevated temperature increase through prolonged pre-evacuation times, given that a case study indicates that the smoke effluent did not spread until late in the incident?
- **Data trace**—charts a variable against time to examine its evolution during the simulation. For instance, does the agent experience expected levels of congestion at specific points during the scenario?
- A **box-plot** presents the bounding values, interquartile values and median for a particular variable. The **box-plot** for simulated and observed data can be compared to provide an indication of the similarity (e.g. overlap) of the results produced. This may be represented in graphical or tabular form.
- A **Subject Matter Expert Test**[6] requires a subject matter expert (independently of the model user themselves) to make a blind comparison between simulated data and a benchmark data-set. This data would be compared to determine whether the subject matter expert is able to consistently and reliably differentiate between them. For instance, is the expert able to distinguish between simulated and real-world arrival curves?

It is possible to apply several of these assessments simultaneously to provide additional insights. For instance, apply plotting data traces and box plots, scatterplots and data traces, etc.

Quantitative means of assessment include the following:

- **Test of difference in event frequency** (e.g. Chi-squared test) establishes whether there are differences between the real-world and simulated frequencies of a particular variable.
- **Correlation analysis** establishes the strength of a relationship between modeled variables.
- **Functional Analysis of Results Convergence (FARC)**—Ronchi et al. [14, 248] developed a method of measuring the convergence of model results with benchmark results, using concepts from functional analysis [269] (based on Peacock et al. [270] and Galea et al. [269, 271]). This employed five measures to determine the similarity between two data traces (i.e. curves) derived from separate resources: total evacuation time, standard deviation of total evacuation times; and from functional analysis, Euclidean Relative Difference (the overall agreement between two curves), Euclidean Projection Coefficient (the best fit between the curves) and Secant Cosine (the differences between the shapes of the curves).
- **Confidence Intervals** requires the comparison of a range of simulated values with a single or a range of real-world observations for the same variable to determine whether there is an acceptable overlap in the results produced [231–234, 236, 238–240].
- **Test of difference in performance measure** (e.g. t-test or Mann-Whitney) is used (given a number of assumptions are met [228–231, 235, 241, 252, 253, 272, 273]) to determine whether a representative real-world and simulated measure examined are significantly different from each other.

There is an overlap between the insights provided by qualitative and quantitative tests. This is apparent from Table 10.3. For instance, a user might want to compare the frequency of exit use. They might do this qualitatively (by charting a histogram to visually compare the results) or more quantitatively (by performing a chi-squared tests on the actual and expected numbers using the routes available, if the statistical assumptions are met).

The user[7] may choose to employ a statistical test of comparison where the required data assumptions are met; however, the use of such techniques is contentious. [237][8] To avoid this issue, and also to enable several of the other tests identified above

---

[6] Might be expanded to the more formal Delphi Method.

[7] Importance of independence within each sample – different runs, summary stat comparison.

[8] Statistical hypothesis tests mentioned above (e.g. chi-square and t-test) typically often employ a null hypothesis that make comparisons based on the assumption that the simulated and benchmark data sets are from the same population – the alternative hypothesis would then state that they are different in some way. As the simulation can only ever be a simplification of the real-world conditions, this null hypothesis is, by definition, false should real-world data be part of the comparison.

(e.g., the FARC method), the user can instead specify a threshold signifying an acceptable maximum difference between the simulated and benchmark data that is then used as a basis for comparison. In such a situation, where statistical tests are not available or deemed inappropriate, a simple user-defined numerical metric could be employed. For instance, consult the *SFPE Engineering Guide for Substantiating a Fire Model* [274] to express the relative difference between simulated and benchmark data.

[simulated.output – benchmark]/benchmark.

This would then be compared against a previously defined acceptable criterion to determine whether the difference in the simulated/benchmark results is tolerable. The user should justify the selection of the acceptance criteria. For instance, the user should examine the variability suggested in other comparable real-world observations to establish a reasonable value for this difference threshold and provide evidence supporting the selected value.

In addition, more qualitative performance measures (observations) can be compared against simulated results (e.g. the paths adopted by evacuees/simulated agents are visually compared) and might be coupled with the previous results to provide greater insight into the model's performance. Irrespective or the metric employed, the user should adopt the following general principles:

- State the nature of the data being compared; e.g. continuous arrival times, location of congestion, exit use, etc.
- State the test being employed and the assumptions on which it is based.
- State the metric used, why it is appropriate, what constitutes success or failure for this metric and why this is acceptable (e.g. within X%, falls within bounds of existing observations, same order of magnitude, occurring adjacent/within the same component).
- State the outcome of the comparison, the implications of this metric value and the user confidence in the stated implications (e.g. simulated data different from real-world data, etc.).

Minimum Level of Comparison

In reality, many of the quantitative forms of assessment may not be viable for comparisons, given that their requirements of the data are not met (e.g. normality, independence, etc.) or there is insufficient data available. At a minimum (given reviewed examples of current practice), the user should report that the following validation tests have been performed:

- *Employed technique CT2 (visual inspection of graphical model output) to examine I5 (paths adopted) and A1 (location of congestion)*
- *Employed technique CT4 (visual inspection of modelled frequencies) to examine S6 (summary of exit use)*
- *Employed technique CT5 (visual inspection of modelled performance measures) to examine I1 (curve of exit arrivals)*
- *Employed technique CT7 (inspection of tabular data) to examine S1 (summary of overall evacuation times)*

However, irrespective of the tests performed, the user should state:

- *What measures are being compared in the test?*
- *What comparative tests are being applied and what assumptions do they make/require?*
- *What acceptance criteria are adopted and why?*
- *What are the resultant differences between the simulated and benchmark data?*
- *What are the implications of these results given the acceptance criteria identified?*

10.4.3 Sensitivity Analysis

Sensitivity analysis is needed to assess the impact of specific modeled factors upon results. This process perturbs the initial conditions to establish the strength and appropriateness of the relationship between initial conditions and model output. This analysis should determine (amongst other things):

- That the **selected** initial model configuration **only** produces the expected aggregate/scenario outcome (i.e. that the user can be confident in the *robustness* of the relationship between the model configuration and the output).

- That **several** distinct initial model configurations **do not produce the same** expected aggregate/scenario outcome (i.e. that the user can be confident in the sensitivity of the model's output to the specific initial conditions represented).

Both cases undermine the validity of the model. In the former case, small changes in the initial conditions do not reliably indicate the expected outcome; in the latter case, other initial conditions (i.e. large changes in the initial conditions) also produce the same expected outcome. This provides an alternative viable model configuration (explanation) of the expected outcome. In both cases, the relationship between initial conditions and outcome is not sufficiently strong.

OUTCOME: Ensure model predictions reflect real-world expectation.

10.5 Reporting Test Results

Given that a model is only ever tested for a specific application/scenario, the user should ideally report the four stages of the modeling process (selection, configuration, verification and validation) when presenting any important results. The absence of a description of these four stages would make it difficult for a third-party reviewer to have sufficient confidence in the modeling process and might potentially reduce confidence in the model user. As noted earlier, in many situations the user may feel that existing testing is sufficient and that they do not need to perform additional tests. If so, they should still address the four stages, but identify either where they have performed new tests or where they rely on previous tests and what these tests are. Any third party reading the report can then make a judgment as to the suitability of the tests performed or the previous tests on which the user relies. During each of the original discussion sections, the key questions for each stage were identified; these are now compiled and represent the *minimum information* that should be provided by the model user. The user should provide evidence as to how (and whether) these questions have been addressed, using their own work or evidence sought from secondary sources (i.e. the net information available). A hypothetical set of answers is provided to illustrate the information that might be expected (for the hypothetical model SFPEMod), although in an abridged form.

Step A. Model Selection—Justify the selection and use of the model.

[MS1]	*What documentation is publicly available informing model selection?*
	– SFPEMod is described in a comprehensive technical guide and is described in five peer-reviewed journal articles (references provided by the user).
[MS3]	*Does the model have sufficient documentation for the user to assess existing model testing?*
	– SFPEMod testing is described in an appendix to the technical guide (supported by example model files to repeat the tests) and by two peer-reviewed journal articles (references provided by the user).
[MS4]	*What aspects of the model are driven by entirely by user actions as opposed to model forecasts?*
	– The user provided data describing the initial conditions (population size, distribution/% with impairment, geometrical structure, exit capacity/availability, travel speeds, flow constraints). The agent response was configured to represent the procedural measures in place: voice alarm notification, guided route use, etc.
	– The rest of the initial conditions were left as model default.
	– The simulation of the subsequent conditions was then determined by the model.
[MS8]	*What techniques are employed by the model to represent key modelling components?*
	– SFPEMod represents the population as a set of individual agents. Each agent is formed from a set of attributes.
	– SFPEMod allows the user to determine individual agent attributes (e.g. pre-evacuation time and travel speed) and behaviors (e.g. route use). Behavioral rules relevant to the scenarios of interest include:
	• Staggering of occupant movement on stairs
	• Reduction in travel speed when using stairs along with tendency to use stair edges.
	• Navigation around other slower moving agents on flat terrain.
	– The performance and interaction of these agents focus on physical aspects, with no reference to social/psychological factors.
	– The geometry is represented using a continuous plane, with agents occupying a coordinate on the plane according to the space available. The plane may represent different types of terrain (corridor, stair, etc.) and flow rates at components are predicted given the terrain and the width of the component involved. The geometry is represented using an architectural diagram.
	– No environmental conditions can be represented.
[MS9]	*What evacuee behaviors are represented by the model?*
	– SFPEMod adopts a rule-based approach. The rules are employed when specific conditions are experienced by the agent. These rules primarily relate to local agent navigation and obstacle avoidance.
	– Agent travel speeds alter according to the terrain traversed (e.g. corridor and stair) and the surrounding population density experienced.
[MS10]	*What output is produced by the model?*
	– SFPEMod produces results at the agent level (evacuation time, congestion experienced, distance travelled) and at the scenario level (overall evacuation time, number of agents using each exit, average congestion experienced, average distance travelled).
	– SFPEMod produces a numerical data file that includes individual attributes (every 30 s) and summary data for performance components, as defined by the user beforehand. It also produces animation files that can be viewed via a VR interface.

Step B. *Model Configuration—Identify the factors being examined, scenarios formed and the model changes required to reflect these factors.*

[MC1]	*What real-world scenarios are being examined and what factors are employed to produce the scenarios of interest?* – The scenario involves the full evacuation of residential mid-rise structure. The structure is represented using an architectural diagram. – It has a mixed population of 200 people. A section of this population will be elderly and/or have movement impairments. – A fire is assumed to develop during the day. – The building has voice notification throughout and route guidance (provided via the notification system and extensive signage) – Two staircases are available (whose positions are the dimensions that are described in the architectural diagram, attached to the report by the user). Both of which are in common use.
[MC2]	*What is the assumed impact of these factors?* – All of the population will be evacuated. – 200 people will be distributed across the living areas, with two people assumed per apartment. – 10% of this population were identified as having a movement impairment. This will adversely affect their travel speed and pre-evacuation time. – Voice notification will influence the pre-evacuation distribution and the route use. – Familiarity with the stairs and route guidance will influence route use.
[MC4]	*What model changes were made to represent the impact of these factors on evacuee performance?* – Population travel speeds: 90% of the population—1.2 to 1.5 m/s; 10% of the population—0.5 to 0.75 m/s [21, 203] – Pre-Evacuation times: 90% of the population—60 to 300 s; 10% of the population—120 to 360 s. [203] – Route Use: 60% of the population—East Stair; 40% of the population—West Stair. Given distribution of population.

Step C. *Model Verification—Confirm that the model changes have been made and the model performs as expected given the new settings. (Also refer to* Table 11-2).

[MVf_I4]	*Do assigned pre-evacuation times lead to agents responding at the prescribed time?* – Report developer tests employing Verif.1.1: Pre-Evacuation Time Distribution, as outlined in Ronchi et al. [248]
[MVf_I8]	*Do assigned travel speeds lead to agents moving at the desired speed given the various terrain types faced (e.g. corridor, stairs, etc.).* – Report developer tests employing Verif.2.1: Speed in a corridor, as outlined in Ronchi et al. [248] – Report developer tests employing Verif.2.2: Speed on stairs, as outlined in Ronchi et al. [248] – Report developer tests employing Verif.2.10: People with movement disabilities, as outlined in Ronchi et al. [248]
[MVf_S3]	*Do imposed flow rates limit the achieved flow rates through the egress component as specified?* – Report developer tests employing Verif.5.1: Congestion, as outlined in Ronchi et al. [248] – Report developer tests employing Verif.5.2: Maximum flow rates, as outlined in Ronchi et al. [248]
[MVf_I5]	*Do agents use assigned routes? Do agents avoid routes that are deemed unavailable?* – Report developer tests employing Verif.2.3: Movement around a corner, as outlined in Ronchi et al. [248] – Report developer tests employing Verif.3.1: Exit Route Allocation, as outlined in Ronchi et al. [248]

Step D. Model Validation—*Confirm that the model's predictions for the scenarios of interest are not invalid. (Also refer to* Tables 10.2 and 10.3).

- What is the source of the benchmark data employed?
  - Observations of a full evacuation from a similar six story residential structure with comparable population size (approximately 250 people) and type (e.g. age range, gender distribution, % of population with an impairment, etc.).
- What is the event scenario from which this data was collected?
  - It was a routine drill. Occupants had no prior awareness of the incident. Environmental conditions did not affect the evacuating population.
- What is the format of the benchmark data?
  - CCTV footage was collected at the entrance to the stairs on each floor and at the final exit points.
  - Estimates of pre-evacuation times, exit use and arrival times were generated.
- What model changes have been made?
  - Pre-evacuation times 30–360 s
  - Population of 250, 25 of whom have a reduced travel speeds (assumed to be 50% of default speed distribution)
  - Stair use according to proximity.
- What measures are being compared in the test?
  - Path adoption, Exit use, Individual/Overall arrival times.
- What acceptance criteria are adopted and why?
  - An acceptance criterion of X% is employed when comparing the simulated model output against the benchmark data.

10.5 Reporting Test Results

- What are the resultant differences between the simulated and benchmark data?

	Subjective Assessment	Application
CT2	Visual inspection of Narrative	*I5—Evidence of Paths Adopted:* Agents used the stair routes, adopted representative paths throughout the building and used handrails on stairs in accordance with footage. (Images provided by the model user) *A1—Location of Congestion:* Congestion was simulated after 1 min at the base of the stairs. The congestion started at the same locations during the drill after 55 s. (Images of congestion provided)
CT4	Visual Inspection of Histogram	*S6—Summary of Exit Use:* There was a 52%:48% split in the exit use during the simulation. There was a 55%:45% split during the drill.
CT5	Visual inspection of Data Trace	*I1—Curve of Exit Arrivals:* Both simulated and drill arrival curves had a similar gradient (X) and had no obvious steps or inflections, indicating that the arrival of the evacuees was constrained by the flow capacity. (Graphs provided by the model user)
CT7	Visual Inspection of BoxPlot/Table	S1- Summary of evacuation times

Drill (s)	Simulated (s)
432	461
[30–630]	[25–663]

- *What are the implications of these results given the acceptance criteria identified?* Given that the underlying dynamics appear representative and the performance measures are within X% of the original incident, these results do not preclude the use of this SFPEMod model for the scenario outlined in B above. In addition, the model developer reported comparison against the RivalMod for mid-rise residential structure *[reference provided by user]* with the results falling within Y% of the RivalMod output for overall evacuation times, given comparable model configuration. This is relevant as RivalMod has been subjected to validation tests against mid- and high-rise residential evacuation drill data.

The following information should then be reflected when test results are reported:

- Egress models simulate scenarios that involve the interaction of numerous factors during an evacuation. The factors represented in the simulated scenarios should be outlined.
- Any egress model is a simplification that reflects limitations in the technology employed, the time available, data, available knowledge, and user expertise. These limitations influence the representation of the real-world factors that affect performance and introduce a level of uncertainty. The simplifications made and their impact on the model output need to be understood and clearly reported.
- Model testing is a key task in determining if the model is appropriate for a task and whether it can produce credible results. The nature, purpose and outcome of these tests should be documented.
- The hypothetical example presented above includes a minimum level of test information that should accompany model use for a given scenario. It is important that a third party reviewing simulated results fully understands the underlying model/scenario assumptions and the demonstrated limitations of the approach adopted—whether it is produced by the model user or produced previously by other interested parties. It is important to reach a basic net level of testing and that this is described in sufficient detail for a third party to make an assessment.

This chapter provides guidance on model testing based on the following four areas:

- Model Selection: Justify the selection and use of the model.
- Model Configuration: Identify the factors being examined, scenarios formed and the model changes required to reflect these factors.
- Model Verification: Confirm that the model changes have been made and the model performs as expected given the new settings.
- Model Validation: Confirm that the model's predictions for the scenarios of interest are not invalid.

A description is provided of the types of actions required to comprehensively address each of these four areas. A set of minimum requirements is provided for each area as an engineering baseline.

Estimation of Uncertainty and Safety Factors

11.1 Introduction

Performance-based human behavior analysis and application to design relies on current scientific knowledge and the ability to perform reasonable, accurate technical predictions with due consideration of uncertainties attributable to the data, assumptions, methodologies or models employed. The purpose of this chapter is to outline considerations and ways to manage the uncertainties in the "people components" of fire safety and allow for more refined analyses with less reliance on safety factors. However, where safety factors are used the guidance in this chapter should be beneficial in determining a basis for the degree or level of safety factor to be applied.

11.2 Sources of Uncertainty

Human behavior analysis and egress modeling inherently has a degree of uncertainty in the conditions represented and the results produced. This uncertainty is discussed in detail by Ronchi [248], Hamins and McGrattan [275], Averill et al. [276] and Notarianni and Parry [314]. In order to manage the uncertainties, the engineer should be aware of the sources of uncertainty to which a human behavior analysis or egress model may be subject. The following are four common types of uncertainty [248]:

- *Measurement uncertainty* is associated with the accuracy of the collection and measurement of real-world data.
- *Parameter uncertainty* is the uncertainty in the values of the parameters of the model. This is often associated with the utilization of real-world data in a configured egress model.
- *Model uncertainty* is associated with the approximations, assumptions and bias (conservative or non-conservative) regarding the representation of human behavior and response during an evacuation.
- *Behavioral or Completeness uncertainty* represents aspects of the real world human behavior that knowingly or unknowingly are not addressed in the model.

Uncertainties when modeling the behaviors of fires and of people can be usefully divided into two categories, epistemic (modeling) uncertainty and aleatory (random) variation. Epistemic uncertainty concerns the validity of our models. The more accurately our models represent reality, the less the epistemic uncertainty. Aleatory variability concerns the randomness or inherent unpredictability of natural phenomenon. Even if our models were perfect, aleatory variability would limit their predictive validity in real world settings. As an example, assume we could perfectly model the effects of ventilation on fire growth, that is, there would be no epistemic uncertainty. In reality, the model would not be able to accurately predict fire growth to the extent that it cannot predict the environmental conditions that affect the fire, including the status of doors, windows, and other sources of air movement—all sources of aleatory variability.

© Society of Fire Protection Engineers 2019
SFPE Society of Fire Protection Engineers, *SFPE Guide to Human Behavior in Fire*, https://doi.org/10.1007/978-3-319-94697-9_11

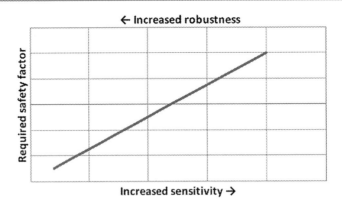

Fig. 11.1 Relationship of Safety Factor Verses Robustness and Sensitivity

11.3 Strategies for Managing Uncertainty

One source that has developed some meaningful and straightforward guidance for managing uncertainties in the fire safety design process can be found in the Nordic publication, INSTA/prTS 950, Technical Specification 950 "*Fire Safety Engineering – Verification of fire safety design in buildings*" [277]. The considerations and strategies to manage uncertainty described in INSTA/prTS 950 are useful and applicable to human behavior analysis and egress modeling and are adapted here as guidance for engineers managing or performing such analyses.

Two basic strategies can be applied to manage the uncertainties in the "people components" of fire safety analysis. These two strategies can be applied independently or in combination to manage uncertainty.

1. Reduce uncertainty by using more accurate data, more thorough analysis or sensitivity analysis, or more complete and robust fire scenarios and robust models (models with validation and verification for the intended scenarios).
2. Reduce the impact of uncertainty by applying a safety factor/safety margin to address the potential known variation in parameters, fire scenarios and/or inherent difficulties associated with the unpredictable aspects of human behavior.

For the purpose of this chapter the following definitions apply:
Safety Factor – A factor applied that expresses the capacity of a system compared to what is needed to reach an objective. Also, a factor applied to a predicted or threshold value to ensure a safety margin is maintained.
Safety Margin – The difference between a predicted value and the actual value where a fault condition is expected.

Figure 11.1 represents the general relationship between analyses using increased robustness verses those that have increased sensitivity and variations in the parameters of the analyses.

It follows from Fig. 11.1 if the safety factor is small, the sensitivity and uncertainty analysis should be very thorough, whereas a greater safety factor should allow for a less comprehensive assessment and have a higher degree of sensitivity in the parameters used. A sensitive design will require a greater safety factor. From this, the assessment of robustness, uncertainty and sensitivity is closely linked to the safety factor that may need to be applied.

Depending on the level of sensitivity, robustness and uncertainty, the need for a safety factor(s) may be increased or decreased based on the following considerations.

Within any evacuation analysis, there should be a safety margin/factor between the Required Safe Egress Time and Available Safe Egress Time. The extent of this should be governed by high level of confidence in the analysis and associated levels of uncertainty. Where there are higher levels of uncertainty then a greater safety margin/factor should be chosen. The selection of a suitable safety margin/factor shall be determined on a case by case project basis.

11.3.1 Reduced Need for Safety Factors

The following methods or circumstances are proposed in order to reduce the need for safety factors when conducting a human behavior analysis:

1. Reliable/Conservative Data
2. Low sensitivity or variance in the parameters
3. Sensitivity Analysis used to evaluate the variance or distribution of parameter values with higher sensitivity
4. Multiple assessment approaches used to identify uncertainty due to theory or methodology
5. Models shown to have completeness, validation and verification appropriate for the analysis are used (untested models are to be avoided).
6. Fire scenarios are robust (credible worst case)
7. Fire safety measures relied upon are capable, reliable, redundant

11.3.2 Increased Need for Safety Factors

The following circumstances are proposed where there is an increased need to use safety factors when conducting a human behavior analysis:

1. Data lacking or does not match well to the context of the analysis
2. Input based on assumptions or without sufficient support
3. Significant outcome variations due to small input variations
4. Little or no sensitivity analysis performed
5. Models can provide results but lack completeness, validation or verification (untested models are to be avoided)
6. Fire scenarios are uncertain
7. Fire safety measures may not be sufficiently capable, reliable, or redundant to be relied on for performance

11.4 Sensitivity Analysis to Reduce Uncertainty

Sensitivity analysis is useful in uncertainty analysis by focusing attention on variables and parameters that have the greatest impact on the results. If a sensitivity analysis can be combined with information on the likely magnitudes of errors in the component values, then a comprehensive calculation of stochastic uncertainty is possible. Sensitivity analyses can be useful in all quantitative analyses to illuminate the robustness of the trial fire safety design.

The objective of a sensitivity analysis is to determine the relationships between the uncertainty in the independent variables ('input') used in an analysis and the uncertainty in the resultant dependent outcomes ('output'). A sensitivity analysis provides information regarding how the variation (uncertainty) in the output of a mathematical model can be apportioned, qualitatively or quantitatively, to different sources of variation in the input of a model. By testing the responsiveness of calculations to variations in different parameters, sensitivity analysis allows the identification of those parameters that are most important to the outcome. It does not necessarily provide information regarding the value that should be used, but it can show the impact of using different values.

A sensitivity analysis may be carried out in the following steps:

1. Specify the parameter or variable and select the inputs of interest. Either one parameter or a combination of parameters made need to be assessed.
2. Select a range of values for the input parameters of interest; ideally, the range ought to encompass the potential limits or bounds of the input parameters.
3. Repeat the model with input values identified in (2) to develop a distribution or range of the output/outcomes.
4. Assess the relative importance of each input factor on the output/outcome.

The following are common parameters that may need to be evaluated using sensitivity analysis when determining the effects of uncertainty with respect to human behavior and evacuation modeling:

- Pre-evacuation time
- Occupant characteristics (e.g., physical condition, age, body size, social relationships, etc.)

- Population density
- Occupant familiarity with surroundings
- Tenability limits
- Travel Speed/Flow
- Route & Path Choice
- Fire Scenarios – modeling or estimation of the fire hazard presents an additional set of parameters that are key to the determination of ASET.

11.5 Robustness to Reduce Uncertainty

The robustness of a human behaviour analysis or evacuation model should be assessed and considered when determining the need and magnitude of a safety factor (if necessary). Depending on the design or scenarios to be addressed, the analysis methodology or model used may range from a simple and straight forward algebraic equation/set of equations, to a complex computational model. Robustness depends not on the complexity or the methodology/model, but rather on the validity and relevance of the methodology/model to the design or human behaviour scenarios to be addressed.

Robustness may also depend on one or more of the following factors:

- The predictive capability and accuracy of a given model or analysis technique (See Chaps. 7, 8 and 9)
- The methodology or model has or has not been validated sufficiently for predicting human behaviour or performance (see Chap. 10)
- The performance of proposed fire safety systems (one or more systems in combination) is or is not sufficiently capable, reliable or redundant
- Are the scenarios being evaluated sufficient to challenge the proposed design (e.g. an exit is disabled or not available for evacuation)

A methodology/model that is well understood, and for which limitations are identified, may well fit the parameters of a design or scenario. In this case, the methodology/model may be deemed sufficiently robust provided the analysis does not exceed the scale of the model or extend beyond the known limitations. However, for buildings with large populations and geometrically large spaces there are many computational models available to more efficiently perform evacuation and movement analysis. These models hope to emulate human behavior; however, the question is how does any one model address the variety of factors that impact human behavior and to what extent are a model's features sufficient and relevant to the design or scenarios in question? In the case of these computational models used to address large populations and complex spaces the determination of robustness relates to the basic question "Does the model contain the features and elements needed for the analysis?"

The following is a list of additional questions to be considered when evaluating and selecting a model. The user of this Guide should refer to Chap. 10 of this guide for a more comprehensive explanation and detailed discussion on model selection, verification and validation.

11.5.1 Evacuation Model Type

Is the model based on optimization, simulation or risk assessment?
Is the type of model suitable for the application?
What are the limitations of the model with respect to the application?

11.5.2 Enclosure Representation

Is the model based on a fine network or a coarse network?
How are different spaces and areas within spaces represented?

How are connections between spaces represented?
How are obstructions within a space represented?
How do these representations influence the model results?
How many nodes, connections and obstructions can the model handle?
How are the data entered to represent spaces, connections and obstructions?

11.5.3 Population Perspective

Does the model use a global or an individual perspective?
If the perspective is global, what general characteristics of the population are represented?
If the perspective is individual, what individual characteristics of the population are represented?
How are the individual or global characteristics of the population entered in the model?

11.5.4 Behavioral Perspective

What type of behavioral perspective does the model employ (none, implicit, rule-based, functional-analogy-based or artificial-intelligence-based)?
How does the model treat people-people interactions and their effects on behavior?
How does the model treat people-enclosure interactions and their effects on behavior?
How does the model treat people-environment interactions and their effects on behavior?
How does the model address physiological factors that influence decision-making?
How does the model address psychological factors that influence decision-making?
How does the model address sociological factors that influence decision-making?

11.5.5 Model Validation

Has the model been validated? If so, how and to what extent?

11.6 Considerations When Using Safety Factors

Safety factors are applied to an analysis to compensate for uncertainties to ensure that the relevant fire safety objectives are met. Estimating the risk of the trial fire safety design is subject to uncertainties, and by adding a safety factor to this estimate there is a margin to the required safety level.

The engineer should be judicious when using safety factors and is best served by using the best models available with input/assumptions based on well researched data and properly defined scenarios. Safety factors can result in a false sense of adequacy of the analysis if the factor is arbitrary and without reasonable basis, meaning that the uncertainty of the analysis is not well understood. Conversely, safety factors can be applied to several variables in an analysis and can compound in such a manner that the end result is overly conservative.

The purpose of the safety factor is to compensate for uncertainties in the analysis. Hence the need for a safety factor will be affected by:

- The method/model
  - If the method/model is less suitable for the actual use, a greater safety factor should be considered. An example is using a model that approximates bulk flow of occupants to all available exits but does not consider detailed flow split and varied distribution to the same exits.
- Reliability of input parameters
  - Reliable sources or test data allow for a reduced safety factor. Uncertainties should be compensated by an increased safety factor or by conservative assumptions

Fig. 11.2 Schematic Description of ASET & RSET with Safety *Margin*

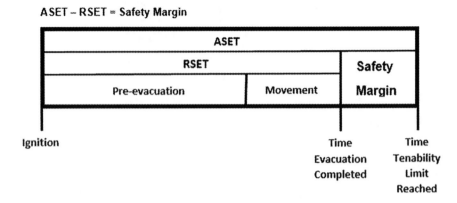

- Sensitivity of input parameters
  - If small changes in input parameters lead to significant changes in the analysis, this should be compensated by an increased safety factor.
- Robustness of the design
  - If the robustness of analyses leads to unacceptable circumstances, a greater factor should be considered, accounting for the capabilities, reliability and redundancies of the actual fire safety systems.
- Difference between a prescriptive solution (e.g. code defined) and the proposed design
  - By definition, the safety factor given by the prescriptive solution is sufficient, as these solutions are defined by authorities or adopted laws. Hence, the safety factor in comparative analyses should address uncertainties in the method or in input parameters.
  - It is important that the comparative analysis must reflect all critical possible differences between the proposed design and the reference building scenario. Safety factors may be irrelevant when uncertainties are proved to affect the pre-accepted solution at least as much as the design alternative.
  - If the proposed design does not differ substantially from the prescriptive solutions, a reduced safety factor could be considered.
- If the input values of all sensitive parameters have been chosen conservatively, thus representing the worst case, no safety factor is needed for these parameters.

Safety factors and/or margins may be applied at multiple points in the analysis process. They may be applied to: (1) selected input, (2) intermediate calculated quantities, or (3) the final outcome as represented in Fig. 11.2 for an ASET-RSET evaluation. The disadvantage of (1) and (2) is that it may be unclear what level of safety has been achieved since the relationship of the input or intermediate quantities to the final calculation outcome may not be linear. While application to (3), the final calculated outcome, provides a clear indication of the level of safety achieved, it is difficult to ascertain how the design might be modified to achieve a different level or margin of safety since the relationship of the input to the outcome is not apparent.

Part III

Fire Situation Management

Enhancing Human Response to Emergency Notification and Messaging

12

12.1 Introduction

In buildings using technology-based warning or messaging systems, it is the goal for occupants to be alerted and provided with critical information in order to respond appropriately. Whether this warning information is provided through audible, tactile, or visual means, the emergency message must be effective so that occupants can act safely and accordingly. In order for a message to be effective in the first place, occupants must pay attention to the message. Since many building occupants may be preoccupied with what they are currently doing, the environment must be changed so that occupants switch their attention from their current activity to the emergency. For example, if the message applies to a movie theater, the projector should be stopped and the lights turned on [11].

As a general note of providing information, attempting to downplay an emergency or using technical jargon to disguise the real situation could confuse people and prevent them from reacting appropriately. It is often a difficult technique to carefully design a successful message. With limited or no information, occupants often do not do what they are expected to do, or even do things that endanger their lives [11]. This acknowledges the importance of creating emergency messages that are effective in real life situations. This chapter provides guidance on how to enhance human response to emergency messages.

Guidance in this chapter can help engineers improve their design of emergency notification and messaging systems, which as a result can help reduce pre-evacuation and therefore total occupant evacuation time. This chapter also provides guidance for building managers, emergency personnel, alarm system manufacturers, codes/standards committees, or others responsible for emergency communication on how to create and disseminate messages using basic communication modes (audible and/or visual technology). Most of the guidance provided in this chapter is taken directly from a report published by the National Institute of Standards and Technology, which was based on a review of 162 literature sources from a variety of social science, human factors and ergonomics, and engineering disciplines [278] and the prioritization of the specific findings extracted from each literature source. This three-year effort was funded by the U.S. Department of Homeland Security, Science and Technology Directorate and the Fire Protection Research Foundation [279].

This chapter first discusses human response to emergency warnings, specifically how people process information and some inhibiting factors that may affect the process. This chapter then presents guidance on how to create and disseminate emergency information in the face of rapid-onset disasters[1] – providing guidance on the dissemination of alert signals, the creation of the warning message, the formatting of messages for both visual and audible means, and the dissemination of warning messages. This chapter then provides emergency communications guidance for buildings with only an alarm signal or those without any alarm system, followed by guidance on emergency drills as a means of message testing.

---

[1] Rapid-onset emergencies are those emergencies that occur with no or almost no (in the case of minutes) notice, rather than slow-onset events (i.e., emergencies in which the occurrence is known hours or even days in advance), and therefore require a different set of emergency messages and dissemination techniques allowing building occupants to receive information in a timely manner, which results in efficient and safer public response.

© Society of Fire Protection Engineers 2019
SFPE Society of Fire Protection Engineers, *SFPE Guide to Human Behavior in Fire*, https://doi.org/10.1007/978-3-319-94697-9_12

12.2 Human Response to Emergency Warning

12.2.1 Processing Information

The Protective Action Decision Model (PADM), discussed in Chap. 4 of this guide, is used as a basis to examine human response to warning systems. The PADM steps are as follows:

1. The individual must perceive or receive the cue(s) (e.g., a visual signal must be seen).
2. The individual must pay attention to the cue(s) (i.e., given that it is possible for the signal to be seen, the occupant actually takes note of the signal).
3. The individual must comprehend the cue(s) and the information that is being conveyed; i.e., given that the signal is noted, that the information is understood.
4. The individual must feel that the incident suggested by the cues and/or information is a credible threat.
5. The individual must personalize the threat (i.e., feel that the incident is a threat to them) and feel that protective action is required (i.e., something needs to be done).
6. The individual searches for what this action might be and establishes options.
7. The options identified are assessed (given the information available) and a final action selected.
8. The individual determines whether the protective action needs to be performed immediately.

As shown above, the individual must go through eight steps before even taking protective action [278]. The PADM shows the steps for how people process information when an emergency message is disseminated. Figure 4.1 simplifies steps 6–8. This chapter later discusses specific guidance on how to enhance the emergency message for effective human response.

12.2.2 Inhibiting Factors

There are several factors that could prevent an individual from following the steps in the PADM accordingly. Some factors that can inhibit human response to warnings include:

- Source-related factors, or factors that originate from the source disseminating the message
- External factors – the physical structure around them and the environmental conditions that they face before and during the incident
- Factors related to the warning receivers, i.e., the population

It is important to note that these factors can inhibit occupants from properly responding to the emergency message. The more effective the emergency message is, the more likely the inhibiting factors can be overcome. See Chap. 4 for additional information on inhibiting factors.

12.3 Guidance on Emergency Communication Strategies

This section provides guidance on the ways in which alerts and warning messages should be created, formatted, and disseminated. The guidance is divided into sections including alerts, warnings, occupants that remain in place, buildings with limited visual and audible notification appliances, and unannounced versus announced drills. It is important to distinguish between the purpose of an alert and a warning message. An alert is meant to grab peoples' attention that an emergency is taking place and there is important information that will be provided to them. The purpose of a warning message is to give that important information to building occupants. Guidance on the construction and dissemination of both alerts and warnings is provided here.

A section on emergency messaging for occupants that remain in place is presented to remind building personnel that emergency evacuation may only be applicable to part of the building. For example, occupants on certain floors in a high-rise building may need to remain in place depending on what floor the fire is on. The following section provides advice for buildings with limited visual and audible notification appliances. Many existing buildings do not have public address systems or mass notification systems and must therefore rely on their trained staff to verbally disseminate the message. The last section discusses the use of unannounced and announced emergency drills, and when it is appropriate to use either. There are certain advantages to using these drills depending on the building's use.

12.3.1 Alerts

It is imperative to disseminate an alert to let building occupants know that a warning message will follow. Regardless of whether the warning message is provided audibly, visually, or via tactile means, an alert is necessary to gain people's attention and should be provided immediately before the warning message.

- Alerts should be significantly different from ambient sounds
- Buildings should reduce background noise when initiating audible alerts
- Flashing, rather than static lights, preferably one standard color for all buildings, can be used to gain attention to visual warning messages
- There are additional methods to alert building occupants to an emergency: disruption of routine activities, tactile methods, social networks, face-to-face
- An alert signal should be accompanied by a clear, consistent, concise, and candid warning message
- If selected, an alert should be tested for its success in getting occupants' attention in the event of an emergency and used as part of building-wide training

12.3.2 Warnings

Warning messages should provide information to the building occupants on the state of the emergency and what they are supposed to do in response to this emergency after an alert signal is given. Warning messages can be provided via visual or audible means. However, before such guidance on message format for visual and audible messages can be provided, it is vital to provide guidance on the content of the warning message itself.

12.3.2.1 The Message

Regardless of the method used to disseminate the warning message, there are certain characteristics that are required of an effective warning message. These are included here:

Message Content
- Research from Mileti and Sorenson [50] on communication of emergency public warnings found that a warning message should contain five important topics to ensure that building occupants have sufficient information to respond. These topics include:
  - Who is providing the message? (i.e., the source of the message)
  - What should people do? (i.e., what actions occupants should take in response to the emergency; for example, whether occupants are to stay in place or not, etc.)
  - When do people need to act? (in rapid-onset events, the "when" is likely to be "immediately")
  - Where is the emergency taking place? (i.e., who needs to act and who does not)
  - Why do people need to act? (including a description of the hazard and its dangers/consequences)
- The source of the message should be someone who is perceived as credible by the building population
- Building managers and emergency personnel should understand the building population and, from this understanding, develop a database of possible trusted sources (as well as backup sources)

The references for each of the following guidance statements can be found in a report published by the National Institute of Standards and Technology [278]. Guidance for message structure, language, and multiple messages is found below:

Message Structure
- Message order for short messages (e.g., 90-characters) should be the following:
  1. Source,
  2. Guidance on what people should do,

3. Hazard (why),
4. Location (where), and
5. Time.
- Message order for longer messages should be:
  1. Source,
  2. Hazard,
  3. Location,
  4. Guidance, and
  5. Time.
- Numbered lists can help to chronologically organize multiple steps in a process
- For limited message length, message writers could draft the message in a bulleted form; each of the five topics in the warning should be separated as its own bullet point
- Distinct audiences should be addressed separately in the message (or multiple messages)

Message Language (or Wording)
- Messages should be written using short, simple words, omitting unnecessary words or phrases
- Messages should be written using active voice, present tense; avoiding hidden verbs
- Messages should be written using short, simple and clear sentences – avoiding double negatives and exceptions to exceptions; main ideas should be placed before exceptions and conditions
- Emergency messages should be written at a sixth grade reading level or lower. An emergency message can be evaluated for its reading level using computer software and/or simple calculation (see Kuligowski and Omori [279] for more details).
- Emergency messages should be written without the use of jargon and false cognates (pairs of words that seem to be cognates because of similar sounds and meaning but have different meanings).

Multiple Messages
- Building managers and emergency personnel should anticipate the need to write more than one emergency message throughout a building disaster, including feedback messages or updates
- In updated messages, tell building occupants *why* the information has changed, so that the new message is viewed as credible
- Provide feedback messages after a "non-event" to inform building occupants that the alert signal and warning system operated and worked as planned and the reasons why the event did not occur
- Building managers and emergency personnel should test emergency messages with the building population

12.3.2.2 Visual Warnings

Messages that are displayed visually will have different capabilities and limitations than those disseminated audibly. Message creators should consider different factors and make different types of decisions based upon the dissemination method. The first consideration is the type of visual technology that will be used to disseminate the messages, which can include textual visual displays, SMS text messages, computer pop-ups, email, internet websites, news (TV broadcast) or streaming broadcast over the web. Depending upon the technology chosen to display visual warning messages, guidance is provided here on message displays to enable occupants to see or notice the displayed warning, understand the warning, perceive warning credibility and risk, and respond appropriately.

Noticing and Reading the Warning
- Place the emergency sign in a location where people will notice it and be able to read it from their original (pre-emergency) location
- Signs will be reliably conspicuous within $15°$ of the direct line of sight
- Text is easier to read when written with a mixture of upper and lower-case letters rather than the use of all capitals
- The recommended relationship for older adults with lower visual acuity is $D = 100 \times h$, providing a more conservative result, and ensuring that a larger population will be able to read the emergency message. In this equation, D is the viewing distance and h is the height of the letter

12.3 Guidance on Emergency Communication Strategies

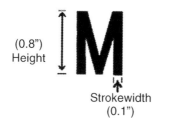

Fig. 12.1 Examples of measuring stroke-to-width ratio [281]

- A stroke-to-width ratio of the letters (ratio of the thickness of the stroke to the height of the letter/number) is suggested as 1:5 (generally), with a ratio of 1:7 suggested for lighter letters on a darker background. Examples of measuring stroke-to-width ratio are provided in Fig. 12.1
- Building managers or emergency personnel should consult the ADA Standards for Accessible Design [280] for additional requirements on signage
- Contrast between the text and the background should be at least 30%, although recommended values could be as high as at least 60%
- The use of pictorials (in lieu of or in addition to text) can also bring attention to the sign
- Message providers should ensure that emergency information is not blocked by other signs or information

Comprehending, Believing and Personalizing the Warning
- Printed text should accompany symbols or pictorials used in visual warnings; a minimum number of words should be used to accompany graphics
- Diagrams that display a series of sequential steps are more successful for comprehension of a process than one single graphic
- Use a color-contrasted word or statement for text that should be read first and/or be perceived as more urgent than the rest, unless color is used for other reasons (e.g., bilingual text)
- A warning message can increase in perceived credibility and risk if occupants are shown that others are also responding
- Simultaneously displayed text (discrete messages) should be used, rather than a sequentially displayed message
- Simultaneously displayed text can also be used for bilingual messages, especially if care is taken to differentiate the text of one language from the text of the other language
- Limit the use of flashing words on visual message displays

12.3.2.3 Audible Warnings

There are specific warning technologies that only (or primarily) affect the aural sense; including public address systems (voice notification systems), automated voice dialling, radio broadcasts, television broadcasts, and tone alert radios. Whereas visual technologies can limit message length, audible warnings are often limited by the attention capabilities of the audience. In other words, an audible message can play for long periods of time with these technology types, and the message creator and source must be careful to provide all important information in an appropriate length of time.

In this section, guidance will be given for methods to increase the likelihood that an individual will perceive, or hear, the message. Then, guidance will be provided that can increase comprehension of the message for audible messages, as well as the ways in which to increase credibility and risk assessment of the event when the warning is presented audibly.

Perception
- Other, non-alert/warning voices in the background should be reduced or eliminated
- Any voice announcements should also be accompanied by simultaneous visual text

Comprehending, Believing and Personalizing the Warning
- Letters are more difficult to identify in speech than numbers, which are more difficult than colors
- Message speakers (or sources) should not be heavily accented and should speak with a rate of approximately 175 words per minute

- Audible warnings should be delivered using a live voice
- Other benefits are provided by a live voice method: messages can be updated with new information and can be used to convey an appropriate level of urgency, if necessary
- Urgency measures should be used selectively to emphasize the more dangerous, immediate, life-threatening situations (since overuse may lead to non-response in future disasters)
- Messages may need to be provided in multiple languages.

Dissemination of the Warning Message
- Use multiple channels to disseminate the warning message – including visual, audible, and tactile means
- A live voice message in an urgent tone is more effective than a prerecorded message, and a familiar voice is best; avoid problems with impromptu phrasing and audibility, (e.g. holding microphone too close and speaking too fast).
- A warning message should be repeated at least once, with some research advocating for message repetition of at least two times
- Messages should be stated in full, and then repeated in full – rather than repeating statements within the same message
- Warning messages should be repeated at intervals, rather than consecutively
- Warning messages should also be disseminated as early as possible
- Face-to-face communication should accompany other audible or visual technologies
- Messages should be disseminated using a combination of both push and pull technologies
- Push communication[2] is most important to use for alert signals as well as initial warning messages
- Staff may need to reinforce messages to encourage occupants to respond

Examples of push communications are text messages and radio broadcasts that interrupt normal programming. An example of a pull communication is a telephone hotline in which individuals must make a concerted effort to seek out additional information without specific prompting. The use of push communications or a combination of push and pull communications provides information to individuals without requiring them to take extra effort to seek it out. These combined message dissemination techniques are effective because they use multiple channels that are by visual, audible and tactile means to alert people about emergencies.

> An example of robust emergency communication is Virginia Tech's (VT) new emergency communication system. Since the crisis that occurred on campus on April 16, 2007 where a student opened fire and killed 33 individuals, the university established a new system that consists of VT phone alerts, VT desktop alerts, digital signage in key academic classrooms and laboratories, broadcast emails to VT email addresses, posts to the VT homepage, outdoor sirens and public address system, and voicemail to VT campus phones [278].

12.3.3 Intelligibility

Voice intelligibility information and guidance is outlined in several references [282], [283]. When designing an emergency notification system, it is important to consider audibility and voice intelligibility to ensure that occupants are receiving and understanding the message correctly. Audibility tests can be used to see if emergency voice messages are at the appropriate sound levels. It is also important to consider conducting intelligibility tests to verify the clarity of voice produced by speakers is adequate. This is often done using an intelligibility meter. Such a meter can be used to measure the potential intelligibility performance of the system. The measurement is either on the Common Intelligibility Scale (CIS) or the Speech Transmission Index (STI).

Note that as with simple sound pressure level measurements, intelligibility measurements at any point will vary. When listening to voice alarm tests, the voice is often understandable even though the measurement may be below what would be considered quantifiably adequate. Also, the voice message is only one aspect of the system. Voice systems are typically

---

[2] Push technologies are those that do not require individuals to take extra effort to receive the alert or warning message (e.g., public address systems or text messages), whereas pull technologies require individuals to seek additional information to acquire the alert/message (e.g., internet websites).

designed to produce multiple cycles of an alert tone to get the attention of the occupants. Also, repeating the alert tone and voice message is important to give occupants an opportunity to focus on the message. Repeating the message allows people to better understand the message or to move toward a speaker to get better intelligibility. The results of the audibility and the intelligibility tests can help the emergency system designer know if they need to improve any audio speakers, sound signals, or any other messaging equipment.

12.3.4 Occupants that Remain in Place

This chapter not only applies to emergency evacuation messages, but to messages for occupants that remain in place as well. Even though occupants that remain in place may be considered to be in 'no immediate danger', they still need to be notified about the emergency, when to act next, etc. The above guidance on emergency communication strategies should be applied to occupants that remain in place, even if the message is delivered verbally by a fire warden, for example. Whether the message is that they should stay tuned for more information or follow specific actions, this information must be disseminated appropriately and effectively.

12.4 Buildings with Limited Visual and Audible Notification Appliances

Many existing buildings are only equipped with an audible alarm signal (e.g. bell, Temporal Code 3, etc.), or no alarm signal at all. The purpose of this section is to provide guidance for such buildings. When only an alarm signal notifies occupants, an alert is disseminated without a warning message. Therefore, it is important that staff, fire wardens, and other emergency coordinators provide face-to-face communication to supplement the alarm. In fact, face-to-face communication may even be more effective since this method has been shown to be more successful than other dissemination methods. Face-to-face communication should follow the guidance on warning message content above as well.

If a building is not equipped with any alerting or warning technology, it is extremely important for fire wardens and/or other staff to alert all occupants and communicate to them the emergency warning information. Emergency coordinators and other appropriate staff should develop emergency plans and train building staff members so that all occupants act accordingly (see Chap. 13, Managing the Movement of Building Occupants). Similar to how it is important to have a credible source provide the message for audible warnings, the same criteria apply for face-to-face communication.

12.5 The Use of Unannounced and Announced Emergency Drills

One way to test the effectiveness of emergency messages is through the performance of fire evacuation drills. In general, emergency egress and relocation drills should be conducted with sufficient frequency to familiarize occupants with the drill procedure.

After each emergency egress or relocation drill, emergency coordinators should produce a written report of the drill to be submitted to designated authorities. The purpose of this step is to document the information and results of the fire drill. The report should include details such as the date, time, participants, location, and results of the drill. This can be a way to test message effectiveness, whether the message is textual, audible, or visual. Message providers should also provide feedback on the results of the drill to the building occupants so that they are aware of how they performed during the drill.

The frequency and methods of the emergency evacuation drills are different based upon the occupancy type of the building [162]. Guidance on the frequency of drills for specific occupancies can be found in the second edition General Guidance on Emergency Communication Strategies for Buildings published by the National Institute of Standards and Technology [279]. The next two sections describe the advantages to conducting unannounced versus announced emergency drills.

12.5.1 Unannounced Drills

There are some advantages to the use of unannounced emergency drills. Building managers, emergency coordinators, and other appropriate personnel should conduct unannounced emergency drills when they would like to observe the building occupants' behavior as those who are not intimate with the fire are likely to behave in a real-life emergency. It is recommended that building personnel conduct unannounced drills in order to observe occupant behavior and assess whether or not their emergency message or technique is effective. The message creator can use observation data to improve alerts, emergency messages, and/or dissemination techniques, if necessary. Providing a debriefing after an unannounced drill is appropriate for people in authority positions to explain the reason for the alarm and summarize the outcome of the drill.

12.5.2 Announced Drills

Though the use of unannounced drills is important when observing occupant behavior, there are some circumstances where announced drills may be necessary. For instance, building occupants in assisted living facilities may benefit from knowledge that a drill is taking place beforehand so that occupants can take preparatory actions. Another example may be in laboratory buildings where operators of experiments might have lengthy or even dangerous shut-down procedures mid-test. Some advantages to announced drills are that they can help familiarize occupants with emergency procedures and prepare them for when a real emergency was to take place. If it is appropriate to conduct both unannounced and announced drills in a building, a good way to observe human response to emergency messaging would be to compare the results from the two types of drills.

Managing the Movement of Building Occupants 13

13.1 Introduction

Whether contributing to a building design or performing a forensic analysis of a building fire, the fire protection engineer can benefit from understanding the on-the-ground considerations for directing people movement during a fire incident in such a building. This understanding should provide insight into whether the building design would be (or was) sufficient in an emergency. This chapter describes two decision models that will help organize the consideration of the large number of factors involved in managing occupant movement [284, 285]. The models organize recommendations from the literature (see Sect. 13.2) that should be considered when making decisions about the movement of building occupants in response to defined fire and occupant behavioral scenarios (see Chap. 6).

The two decision models are closely related.

- The first "design" model (Sect. 13.5) assists the design team in selecting occupant movement strategies well-suited to responding to specific scenarios. The design model can influence both the fire-related procedures and the design features of buildings.
- The second "operational" model (Sect. 13.6) provides guidance for engineers who need to consider whether to add fire safety features and/or change occupant movement strategies in buildings that are already occupied. The second model takes into consideration that various problems may require operational plans to rescue occupants who are unable to escape potential exposures to heat and fire effluents.

In both cases, the models divide occupants into groups that are asked to follow specific movement strategies, depending on the scenario under consideration.

In both models, it is recognized that conditions are evolving during the course of the fire. For example, smoke and/or flames might fill egress paths that were initially clear. In addition, firefighting and rescue operations might affect the speed at which occupants can proceed. Thus, both models include the need for periodic updated communications, either to inform occupants and responders of the changes or to confirm that no change has yet occurred.

The models, in themselves, will not provide definitive decisions about how the movement of occupants should be managed. They will, however, present a means to consider the numerous factors that play roles in the use of evacuation, relocation and protect-in-place strategies when deciding how building occupants can best be safeguarded during fire emergencies. The models are represented as flow charts that show the decision processes that lead to recommendations about which occupants should remain where they are already located, and which occupants should move to safer locations. While this chapter is directed towards the needs of fire protection engineers, it should be of value to building designers and people who plan for and manage the operational movement of occupants during fires.

> The decision-making process of occupants as opposed to designers and managers is covered elsewhere in this Guide. See Chap. 4, Occupant Behavior Concepts – Cues, Decisions & Actions.

© Society of Fire Protection Engineers 2019
SFPE Society of Fire Protection Engineers, *SFPE Guide to Human Behavior in Fire*, https://doi.org/10.1007/978-3-319-94697-9_13

13.1.1 Persons Responsible for the Design of Buildings Before They Are Occupied

Fire protection engineers, architects, building owners, and managers can use this model to clarify the occupant movement strategies that the building is designed to support, and subsequently to provide valuable information to persons who will eventually occupy the building about the fire protection features that support those strategies.

> While not well documented, there is evidence that persons charged with the responsibility of managing a fire emergency have sometimes made serious mistakes. Chertkoff and Kushingian [286] reviewed documentation for several multi-fatality fires, and concluded that managers postponed responses and made poor response decisions that probably contributed to large losses of life. For example, managers delayed ordering an evacuation because they feared that occupants would "panic." In many instances, staff members were unavailable to direct occupants to safe egress routes. Chertkoff and Kushingian also explain that managers failed to comprehend that preconditions were likely to impede an effective evacuation response. These included highly combustible interior finishes and convoluted and blocked egress routes.

13.1.2 Persons Responsible for the Operational Management of Occupant Movement After Buildings Are Occupied

Occupant movement managers plan for and/or direct building occupants during building fire emergencies. Persons who assume the role of occupant movement manager can include building management, building emergency team members, and responding fire fighters. The model can be used to both plan for an emergency and adapt the plan depending on the real-time developments as an emergency evolves. See Sect. 13.10 for a discussion of how maintaining good situation awareness together with applying the model can be used by occupant movement managers to adapt to unforeseen developments during fire emergencies).

Not every building has someone onsite to manage a fire emergency, but the model will also help managers and fire departments develop appropriate training and educational materials tailored to the occupant movement strategies used in specific buildings.

The model for occupant movement managers differs somewhat from the model for building designers: It takes into consideration the possibility that plans may need to compensate for inadequacies in building designs and fire protection features in existing buildings or in buildings where design, maintenance, or operational errors cause unanticipated problems. This decision model helps occupant movement managers understand how strategic responses to fires can take advantage of building features and human capabilities under a range of scenarios. The model can provide guidance both for planning people movement strategies, and for adapting those strategies during a fire emergency.

The use of the model for design is presented first (Sect. 13.5), followed by the model for the operational management of occupant movement (Sect. 13.6).

> The term "movement" is used instead of "evacuation" to clarify that strategies other than leaving the building may be involved, including sheltering-in-place and relocating to a safer location in the same building.

13.2 Available Resources for Tailoring Occupant Movement Strategies to Specific Buildings

There are references that provide valuable information about the various factors that should be considered when planning how to manage occupant movement during fire emergencies. However, these references provide little help in deciding which occupants should be located where given a specific scenario. Examples of literature that provides general advice include the *Life Safety Code*® [162] where detailed recommendations are available for conducting required life safety assessment in assembly occupancies. Burtles [287] has published a guide based on the process of business continuity planning. In England

and Wales, occupancy-specific "fire safety documents" are available online [288]. Moreover, extensive coverage is available regarding the physical design of building features that enable the safe movement of occupants [131, 289]. There is a considerable body of research concerning the modeling of occupant movement during building evacuations (see Chaps. 9 and 10). While these references provide valuable information, they do not explain how to plan for and execute occupant movement strategies that are tailored to specific scenarios in specific buildings.

General guidance about how to assemble options to create life safety strategies are not necessarily tailored to specific fire scenarios in specific buildings. Occupant movement strategies that may be overly simplistic in large and complex buildings are often categorized as follows [289, 290]:

- Simultaneous whole building evacuations. All occupants leave the building at the same time when they are notified.
- Protect-in-place. Also called defend-in-place and shelter-in-place, occupants remain where they are already located throughout the event.
- Phased whole building evacuations. All occupants leave the building, but in a phased sequence based on the vulnerability and/or proximity of building occupants to the fire
- Partial building evacuations. All occupants in a certain part of the building leave.
- Relocating occupants within a building. Persons in the building relocate to safer areas.

More realistically, a combination of strategies may be required for different groups of occupants depending on the informational inputs for the questions in the decision models described below. For example, in a tall office building, persons below the fire zone may be requested to evacuate the building (a partial building evacuation), persons in the fire zone may be requested to move below or above the fire zone (relocation), and persons with disabilities may be requested to move to refuge areas to await rescue (defend-in-place). However, given the higher level of compartmentation and the delayed responses likely from residents in a tall *residential* building, the decision model may lead managers to recommend an entirely different set of strategies—occupants who are on the fire floor may be asked to relocate to a lower floor (relocation), while occupants on other floors may be asked to remain in their rooms or apartments (defend-in-place) [291]. Instead of recommending a general strategy for a building, it often makes more sense to divide the building occupants into groups that should use different movement strategies.

13.3 Factors and Assumptions Used to Divide Occupants into Groups that Require Different Movement Strategies

The models dealing with building design and operational management both require dividing building occupants into groups based on the different strategies that should be recommended to them during a particular fire and occupant behavioral scenario (See Chap. 6, Development and Selection of Occupant Behavioral Scenarios). The data used to divide building occupants into groups will vary according to the selected scenario. For example, the growth and mitigation of a fire depends on the scenario, which in turn, will determine which building occupants are in danger and which are not.

Groups are identified based on five types of information:

- The locations of building occupants (Sect. 13.7.1);
- The anticipated growth of hazardous conditions, based both on the size of the fire and on fire protection features of the building (Sect. 13.7.2)
- Separations from hazards when occupants are stationary, either in their current locations or after they move to the safer location that are their destinations (Sect. 13.7.3);
- Separations from hazards when occupants are moving to safer locations (Sect. 13.7.4);
- Limitations in abilities of building occupants to move to new safer locations (Sect. 13.8);
- The availability of assistance to compensate for those limitations (Sect. 13.9).

The ways that occupants are grouped may differ, depending on each the following assumptions:

- In planning for a fire emergency, the optimal decision for any particular group of occupants depends on the fire scenario. The specification of specific scenarios is generally required for performance-based design solutions and includes both building and occupant characteristics [290]. For the purposes of performance-based design, scenarios generally require quantification. However, for the purposes described in this chapter, a qualitative analysis may be sufficient. The design and operational models specify informational inputs needed to make decisions about where groups of occupants should be located. Each of the informational inputs in the decision model will vary depending on the scenario. (A discussion of scenarios is found elsewhere in this guide. See Chap. 6, Development and Selection of Occupant Behavioral Scenarios).

- For the persons in any particular group, the occupant movement manager recommends only one of two alternatives: either people remain where they are already located or they move to another specific location. If people are asked to move, the route is recommended and assistance is provided to the extent that it is needed.
- Dividing occupants into groups does not depend solely on occupants' locations relative to the fire. Occupant capabilities as determined by mobility, sensory and cognitive limitations must also be considered. For example, occupants in a single location may fall into two different groups, those who can leave the building unassisted, and those with disabilities who should remain in place (and potentially wait for rescue as a backup strategy) because reliable means for them to quickly leave the building are not immediately available.
- Due to uncertainties, the optimal decision for different groups of occupants can change during a fire emergency. Therefore, the occupant movement manager should maintain sufficient situation awareness so that an appropriate backup strategy can be implemented. For example, in the event that smoke migrates to areas that were thought to be well-separated from the fire, the occupant movement manager may need to recommend that affected occupants relocate to safer spaces. For more information about adapting plans during fire emergencies, see Sect. 13.10.

13.4 Delayed Movement for Persons with Critical Functions

When planning for and executing people movement during fire emergencies, it can be important to acknowledge that not everyone is expected to immediately follow the recommended strategies for moving occupants. An important group of persons are those with critical responsibilities that may delay their movement. Regardless of their added responsibilities, people should always be advised to leave before they are in danger. Categories of people in this group may include the following:

- Building emergency response team members, such as floor wardens, typically leave an area later than the occupants for which they are responsible. However, response team members should be cautioned to leave anyway when encountering uncooperative occupants. Areas of a building may be searched to ensure that occupants have complied with the recommended procedures before members of the team leave. Other persons may be delayed because they are responsible for assisting persons with disabilities. The persons managing occupant movement and monitoring the status of fire protection systems may remain in a fire command station to improve the situation awareness of arriving public safety personnel and to cooperatively continue their activities as the emergency develops.
- Persons responsible for securing key tenant infrastructure may be delayed. Examples include persons who save working files and shut down and secure computer systems. As with building emergency response team members, facilities managers may be requested to report to the fire command station to assist firefighters in understanding how to operate building equipment such as heating, ventilation and air conditioning (HVAC) systems.
- Persons responsible for shutting down key building infrastructure, such as manufacturing processes and gas supplies.
- Persons responsible for removing a roster or register of persons who were in the building at the time of the emergency. In many buildings, it is unrealistic to expect an accurate roster of building occupants, especially visitors. However, some buildings have access control devices that track when people both enter and leave the building and a register showing when visitors arrive and leave the premises.
- Persons responsible for maintaining access control during the emergency [292]. For example, security personnel may be located at points of building ingress into the building to prevent entry.

People in any of the above categories that are found in the building should be incorporated in occupant movement planning.

13.5 The Model for Designing Buildings that Optimize Decisions About the Movement of Building Occupants

The decision model that provides guidance to building designers is shown in Fig. 13.1. This is the first of the two models; the second model, concerning the operational planning of people movement strategies in existing buildings, is discussed in Sect. 13.6.

13.5.1 Ways that Designer Can Use the Model

This first of the two decision models is useful for fire protection engineers and other persons who design buildings in four ways:

1. The decision model can guide the inclusion of design features that facilitate better strategies for protecting building occupants. During the design of buildings, the occupant movement decisions that need to be made during a fire emergency should be considered. This includes the fact that responder activity might affect occupant movement speed and/or choice of egress path. The model can be used as a basis for discussions between fire safety professionals and persons who are less sophisticated about the fire protection features that are being considered for a building. Buildings can sometimes be designed to enable improved decision making during fires and to ensure that designs are consistent with the behavioral tendencies of occupants.

    The intended use of this chapter is similar, although specifically focused on occupant movement, to the recommendations for the use of NFPA 550. This Standard explains that the "Fire Safety Concepts Tree can be used as a design tool." Section 13.5.2 explains that "perhaps the most important use of the tree is for communication with architects and other professional involved in building design and management." [293]

2. To the extent that occupants follow recommended actions from building movement managers, the uncertainty about how people are likely to behave during fire emergencies is likely to be reduced. During a fire emergency, the behaviors of individuals will vary depending on the type of information available to them. Information from persons managing occupant movement is an important source of information that can strongly influence behaviors, thus reducing the variability of behaviors and the amount of time that people take to settle on an appropriate course of actions [294]. Quicker and less variable occupant responses should reduce the Required Safe Egress Time (RSET), thereby increasing the margin of safety provided by the design of the building (Available Safe Egress Time or ASET). (For more information about designing messages, see Chap. 12.

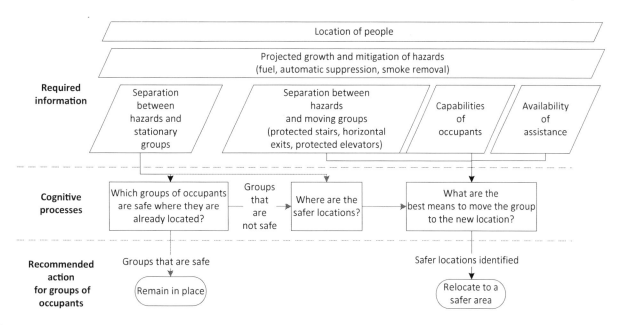

Fig. 13.1 Flow Chart for Designing Buildings that Optimize Decision for the Movement of Occupants

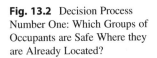

Fig. 13.2 Decision Process Number One: Which Groups of Occupants are Safe Where they are Already Located?

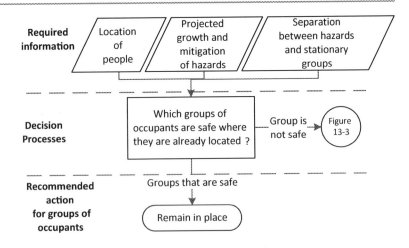

> As an example of how the model can be used to facilitate communications during the design of a building, the emergency movement of people leaving a day care center in an office building requires special attention. The building design ideally facilitates the decisions of persons managing the movement of children in the day care center by providing an easily navigated path to safety. The task can be difficult for two reasons: (1) it can be difficult supervising children when shared routes are crowded with other occupants; and (2) parents in the same building may want to travel to the day care center to ensure the safety of their children creating counter flow and requiring staff to account to children who have been separated from groups by their parents. These difficulties can be greatly reduced by a building that supports a sensible people moving strategy, perhaps by providing day care centers with direct exits to the exterior of the building.

3. The decision model can guide the inclusion of building features that enhance the situational awareness of people who manage the movement of people during fire emergencies. Better situation awareness enables occupant movement managers to adapt their recommended strategies depending on the situation as it evolves during the emergency. For example, a fire protection engineer might recommend that a CCTV system include cameras in stairs and monitors in a fire command station that enable persons managing the movement of people to detect problems more quickly. (Also, see Sect. 13.10).

4. The decision model can help harmonize building design with decision making during fires. The fire protection engineer can provide a "user's manual" that explains how occupant movement managers should use the fire safety features of the building when planning for and executing emergency action plans. Once the building is occupied, building managers should understand the fire protection features, and limitations, of the building—especially as regards their decisions about whether and where to move which people.

> As an example of the importance of harmonizing codes and occupant's movement strategies, codes require that individual patient rooms in nursing homes and hospitals are separated from other areas by fire and smoke barriers. As a generalization, patients in nursing homes can be moved only with considerable difficulty given the limitations of patients and the number of staff. However, doors to patient rooms are normally kept open so that the staff can more easily view the status of patients. Despite the fire protection features mandated by codes, patients have died in nursing home fires because the staff failed to immediately close patient room doors [295]. Considerable effort is devoted to remedy the situation by training staff to close the doors to patient room early during any fire emergency. Unless the doors to patients' rooms have been closed, the used of compartmentalized patient rooms has little value. While harmonizing building fire protection features and the management of building occupants is less obvious in other occupancies, the movement of people still needs to be consistent with the fire protection features designed into the building.

As shown in Fig. 13.1, the building design decision model involves a sequence of three decision processes applied to a specific scenario. The decision model also describes the informational inputs needed to apply the decision processes so that it is clear which movement strategies should be recommended to specific groups. The decisions and recommendations for the movement of occupant groups differ somewhat between the *design* model presented in this section, and the model for the *operational management* of occupant movement presented in Sect. 13.5. However, in both the design model, and the operational model the informational inputs remain the same. Therefore, to avoid unnecessary redundancy, informational inputs for both models are discussed in Sect. 13.7 following the description of the second model.

13.5.2 Decision Process One: Which Groups Are Safe Where They Are Already Located?

This is the first decision for an obvious reason; if occupants are very likely to remain safe throughout the fire emergency where they are already located, then it is simplest to instruct them to stay where they are. This also frees the people managing the movement of occupants to focus their attentions on the remaining two decisions. As described by Bukowski and Tubbs [290], "protect-in-place strategies also known as defend-in-place or shelter-in-place…involve providing adequate safety features to allow occupants to remain-in-place during the event."

When occupants are likely to remain safe in their present locations, they should be instructed to remain where they are already located. Because of personal observations (for example, of arriving fire fighters and smoke) and the use of social media (for example, texting and email from persons outside and in other parts of the building), occupants who are expected to remain in their current locations may not reliably do so in the absence of instructions. Bukowski and Tubbs [290] explain: "Occupants may become aware through communication technology, such as texting, and social media, rather than building notification system." The characteristics of effective building communications systems and messaging are covered in Chap. 12, Enhancing Human Response to Emergency Notification and Messaging. Nonetheless, maintaining good situation awareness is important because fire growth and spread of fires is difficult to accurately predict. Even when occupants are judged to be safe in their current location, it is important to periodically check whether conditions remain relatively smoke-free during the duration of a fire and to reassure occupants in safe areas. As discussed later in Sect. 13.10, where conditions deteriorate in a space, the sequence of decisions must be revisited to determine the next best course of action.

13.5.3 Decision Process Two: Where Are the Safer Locations?

When designing a building, a "safer" location needs to be designated whenever occupants may not remain safe where they are already located. The detail from the design model (Fig. 13.1) is presented in Fig. 13.3. Bukowski and Tubbs [290] explain; "relocating occupants from an area of potential hazard to a protected area of refuge or other safe place within a building can be a safe and effective strategy. The decision about whether there is a "safer" location is discussed in the model for the operational management of occupants in Sect. 13.6.

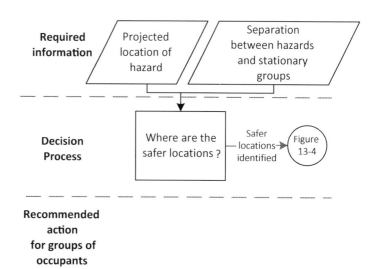

Fig. 13.3 Where are the Safer Locations?

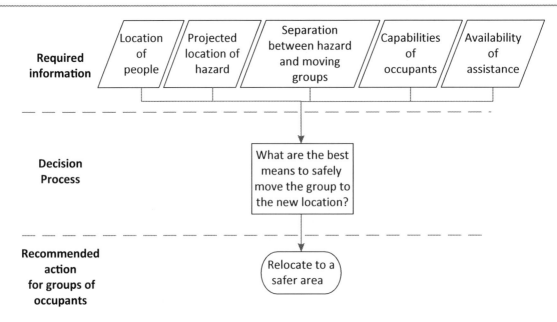

Fig. 13.4 What are the Best Means to Move the Group to the New Location?

13.5.3.1 The Projected Location and Extent of the Fire Hazard

The same basic considerations are pertinent to all the decision processes in the model. In particular, this requires an assessment that the paths of travel to a safer location are likely to remain relatively free of smoke and heat. This will depend on the fire protection features of the building.

13.5.3.2 Are Occupants Safer Outside the Building?

As a generalization, occupants are safer outside of the building where there is a fire. There are exceptions where occupants should be advised to remain inside the building or avoid certain exit discharges. Examples include fires outside of building, small contained fires coupled with dangerous outside weather (e.g., extreme cold, tornados, fires that involve the exterior of the building, and fires and other hazards (e.g., earthquakes) that create falling debris, such as breaking windows. When a fire is located outside the building, smoke often enters the building leading to the misunderstanding that there is an interior fire. Poor situation awareness may lead building an occupant movement manager to evacuate all or part of the building, thinking that occupants will be safer outside. Accordingly, an accurate understanding of location of the fire and other hazards are needed to assess which, if any, areas outside the building are safer than inside.

13.5.4 Decision Number Three: What Are the Means to Relocate Occupants to the Safer Location?

The decision about whether to move occupants to a safer location depends on whether they can be safely moved without exposing them to hazardous amounts of heat or smoke. Determining the safety of routes and providing that assistance needed to use those routes is the most complex of the decisions facing occupant movement managers (Fig. 13.4).

As with all of the decisions, the locations of occupants and hazards needs to be determined before the question about whether safe means to move occupants is answered.

13.6 The Decision Model for the Operational Planning of People Movement During Fire Emergencies

As indicated in Chap. 4, during the process of taking protective actions in a building fire, building occupants make decisions about how to respond to the often-ambiguous information available during many fire emergencies. Ideally, the most valuable sources of information during many emergencies are the persons responsible for recommending a best course of action to

13.6 The Decision Model for the Operational Planning of People Movement During Fire...

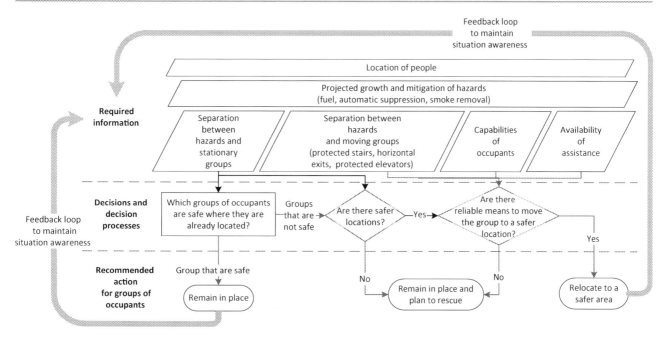

Fig. 13.5 Decision Model for Operational Managers of Occupant Movement

building occupants (See Chap. 12). These may be people employed by the building or fire fighters responding to the emergency.

This model for the operational management of occupant movement, represented in Fig. 13.5, differs somewhat from the model for building designers; it takes into consideration potential limitations in occupied buildings that must be compensated for when deciding which occupants should move where during various fire scenarios. The express use of NFPA 550 is similar; Sect. 13.5.4 states that "The Fire Safety Concepts Tree can be used to assess the fire safety in an existing building." Where the generally of operational managers extend beyond the "operational" model described in this section, NFPA 550 can be used as a guide to the entire range of factors affecting life and property protection in an existing building [293].

Occupant movement managers may inherit problems in building design that result from a variety of sources:

- The building may have been designed using code provisions that do not meet current standards for fire protection,
- The fire protection features incorporated into the building have not been maintained and may fail to function as designed.
- Occupancy or site conditions may exist that were not considered during the design of the building.

The operational model for occupant movement management considers the possibility that building occupants may need to be rescued. This might occur either because:

- A group of occupants may not remain safe during the duration of the fire but safer locations are not available; or,
- There is a safer location than where a group of occupants is already located, but the means to safely relocate the occupants is unavailable.

The operational model differs from the design model described in Sects. 13.5 and 13.6 in the following ways:

- Instead of asking "Where are the safer locations?" the question is asked "Are there safer locations?"
- Instead of asking" "What are the means to move the group to safer location, the question is asked "Are there reliable means to move the group to a safer location?"

In either case, the answer may be "no," and the best alternative is to leave the group in their present location and plan to rescue the occupants. Of course, in a well-designed building that meets current fire protection standards and when fire scenarios are expected, the need to rescue building occupants should not occur.

The model also includes a feedback loop indicating that conditions should be periodically reassessed to determine whether occupants remain as safe as originally expected (see Fig. 13.5).

13.7 Informational Inputs Common to Both Versions of the Model. The Number of Occupants in the Various Locations of the Building

In both the design and operational models, the same informational inputs are required to make actionable decisions about which building occupants should be located in which areas. These common inputs are discussed in this section.

13.7.1 Locations of Building Occupants

Given a scenario, decision-makers need a reasonable accurate idea about where occupants are likely to be located in the building. At the onset of an emergency, the most conservative answer is usually that there may be people in any and all occupied spaces. In certain circumstances, for example, during maintenance or retrofit work, there may be people in spaces normally considered as not occupied. Further, in some occupancies, the numbers and locations of building occupants varies considerably depending on working or event schedules.

13.7.2 Projected Growth/Mitigation of Fire Hazards

When deciding whether occupants are safe in their present locations, there needs to be credible expectations or estimates of the location and extent of the fire hazard. A set of credible challenging scenarios considers the size and location of the fires (see Chap. 6). The mitigation of fire hazards is also a function of design features such as automatic suppression systems and smoke removal systems. When designing complex buildings, it often is useful to model the growth of design fires and the extent that the associated hazards develop and spread in the building. More detailed and technical discussions these building features can be found in Bukowski and Tubbs [290] and in Tubbs and Meacham [289].

13.7.3 Building Features that Separate Stationary Occupants from Fire Hazards

This section describes building features that maintain separation between occupants and fire hazards while they remain in a particular location. The separated location can be where they are already located at the onset of the emergency, or their destination if they are requested to move to a safer location. Building features are generally prescribed by the requirements of building codes and are reviewed here only to highlight the types of information that people movement strategists should understand when devising their strategies. More detailed and technical discussions these building features can be found in Bukowski and Tubbs [290] and in Tubbs and Meacham [289].

13.7.3.1 Exits

There are two basic types of exits, horizontal (e.g. interior door assemblies) and vertical, usually stairs. Occupant movement managers can direct occupants to relocate to exits where they are relatively safer than the spaces that they are leaving. However, exit stairs are rarely recommended as the final destinations during fire emergencies for two reasons. First, fire hazards can still enter the exits (especially because fire fighters often use exit stairs to stage their fire suppression activities, allowing smoke to enter). Second, exits can become so crowded in some large buildings (e.g., commercial high rise and assembly occupancies) that some occupants may be unable to enter the exits when the entire building is evacuated at the same time. Occupant movement managers should understand the value and limitations when exit stairs and horizontal exits are used as destinations and consider how occupants can be rescued if necessary.

13.7.3.2 Separated Areas

Separated areas provide protection inside buildings and are designed to provide safe destinations during a fire emergency. In addition to fire and smoke barriers, separated areas may potentially include a means to protect the areas from smoke, for example, a pressurization system that provides air from the exterior of the building and that impedes smoke from entering the area; a two-way communication system that helps the manager maintain situation awareness; and, direct access to an exit.

Situation awareness means maintaining awareness of the numbers of occupants taking refuge and likelihood that it would remain protected from fire hazards (see the Sect. 13.10.4, Definition of and Importance of Situation Awareness while Managing Occupant Movement during Fire Emergencies).

> The term "separated areas" is used instead of "refuge areas" and "compartment." The definitions of "refuge area" and "compartments" are technical in nature and may vary by codes provisions. In the context of this chapter, separated areas are defined as areas that are relatively safe from fire hazards by the use of fire and smoke barriers.

In many buildings, separated areas are primarily intended for use by persons who are unable to evacuate without assistance to the outside of a building because they have disabilities or are patients in a health-care facility. In some specially designed buildings, separated areas may be intended for all occupants. For example, some very tall buildings have dedicated entire floors as refuge areas so that occupants do not need to descend through the entire building to reach relative safety [296]. Such refuge floors, appear to offer a compelling way to create an entire floor that will function as an area of refuge within a building where a large number of building occupants can be gathered. However, many questions arise about whether a refuge floor provides an increased or a decreased level of safety, and whether there are more effective and efficient ways to accomplish the goals that refuge floors attempt to achieve [290, 297].

Some buildings have separated areas that may be referred to as "areas of rescue assistance," meaning that occupants may not remain safe during the duration of a fire, but can remain in relative safety pending the arrival of assistance to move to even safer areas (see Sect. 13.10). These areas are better considered as areas of rescue assistance where occupants who wait temporarily until they can be safely evacuated by fire fighters. As an example, many building have enlarged spaces in exit stairs that can accommodate a few wheelchair users while other persons use the exit stairs to relocate to safer areas inside or outside of the building.

When the fire hazard is not modeled using a prescribed engineering method, reasonable judgments are still possible based on building features and the likelihood that occupants in various locations will remain separated from fire hazards. While persons well-educated in fire protection methods have little difficulty with these judgments, persons charged with the responsibility of the operational management of moving of building occupants may not accurately understand how building features affect fire growth and spread. Persons who design the fire protection features can look for opportunities to provide written guidance about the building features that keep occupants separated from a fire hazard, including, at a minimum, the features covered in this section and in Sect. 13.7.4.

13.7.4 Building Features that Separate Moving Occupants from Fire Hazards

This section describes building features that maintain separation between occupants while they are moving to a safer location. More detailed and technical discussions these building features can be found in Bukowski and Tubbs [290] and in Tubbs and Meacham [289].

13.7.4.1 Stairs

There are two basic types of stairs, "exit" stairs and all other types of stairs. Other types of stairs will be less well-protected. At a minimum, exit stairs are protected by fire-rated barriers and closed doors. Therefore, occupants are likely to be relatively safer while relocating to areas where they are better protected from fire hazards.

Stairs that do not qualify as part of an exit may be an acceptable means to relocate occupants, but they should only be used when (1) exit stairs are unavailable, and (2) when they have been examined and found to be relatively free of smoke.

Apart from single and two-family homes, multistory buildings, with rare exceptions have at least two stairs. Older residential structures may have only a single "exit" stair, preferably with a fire escape serving as the alternative means to escape from upper floors.

Stairs, including exits, can be compromised by firefighting operations and stack effects in tall buildings. Additional protection for exit stairs is provided in certain buildings where the exit stairs are protected using either smoke enclosures or pressurization. Occupant movement managers should have a basic understanding of how these features provide additional protection and in what ways they are limited.

13.7.4.2 Protected Elevators

For the most part, the use of elevators during fire emergencies has been prohibited. However, with additional features such as protected lobbies, elevators that are relatively safe are increasingly being used to move occupants during fires.

Elevators are always taken out of service when smoke is detected in elevator lobbies, hoistways, and machine rooms. When elevators are taken out of service (elevator capture or phase I service), they travel to the ground floor (or another floor if smoke is detected on the ground floor) where the doors open and the elevator become inoperable. After elevators are taken out of service, they can still be used by fire fighters using a special key (firefighter service or phase II operation). Fire fighters may use elevators during a fire emergency to move equipment closer to a fire and to evacuate persons who have been unable to evacuate without their assistance—but only after their safe use has been carefully evaluated.

In recent years, there have been efforts to design elevators that can be used to move occupants to safer areas during fire emergencies [290, 298, 299]. These elevators, and the lobbies where building occupants wait for elevators to arrive, are likely to be pressurized and enclosed with smoke and fire-resistant barriers. Two-way communication with the fire command post and signage are designed to assist wardens and occupants in deciding whether to use the elevators instead of exit stairs. Direct access to exit stairs from the elevator lobby provides a backup strategy in the event that a protected elevator becomes unavailable. Further, the elevators are designed to be taken out of service when smoke is detected in the lobbies or hoistways. Very careful planning by occupant movement managers and firefighters is essential to ensure that such elevators are used effectively to move occupants to safer locations. They should have detailed knowledge of the automated functions of the elevator system. This will enable them to monitor the routing of elevators to ensure that they are operating as designed and that they are not manually removed from service while still operating effectively to relocate occupants. Occupants need to be educated about the features of protected elevators that enable their use during fire emergencies.

13.7.4.3 Exterior Supplemental Evacuation Equipment

Recent innovations, primarily chutes and moving platforms, can be installed and deployed on the exterior of buildings. In general, building and fire codes do not qualify these systems as acceptable means to escape buildings because the systems may not be reliable and can only move small numbers of occupants.

13.7.4.4 Controlled Descent Devices

Controlled decent devices are used to move persons with disabilities or injuries that prevent them from using stairs. These devices, often referred to as evacuation chairs, use a track system that allows them to be moved down stairs by a single operator. Research on the demands on device operators has taken place [300], and performance standards have been developed for the track type evacuation chairs [301]. Using a controlled descent device requires that the occupant be transferred to the chair, an operation that can injure the occupant or operator when poorly executed. Practice in using the devices is recommended to provide both operators and occupants with the confidence to use these devices safely. Persons with disabilities should be consulted about how they can be most safely transferred to and from the devices. Evacuation planning should also include the identification of any equipment necessary for horizontal travel at the floor of discharge (e.g., continued use of controlled descent device, use of manual wheelchairs).

13.8 Limitations of Occupants

This topic is also covered in Chap. 3 of this guide. An assessment of occupant capabilities is essential when determining whether there are means available to move occupants to safer locations. Certain building occupants are likely to be less capable, or entirely unable, to move to other locations in the building without special consideration. This is especially likely when occupant characteristics limit their abilities to use stairs. Occupants may have limitation because of their age (either very young or old), mobility, sensory and cognitive disabilities (many of which may not be apparent to an observer), and temporary "disabilities" such as pregnancies and injuries. It is important not to categorize persons without regard to the effects that their disabilities have on the abilities to respond with assistance. In planning for occupant movement, persons should be encouraged to participate to the extent that their impairments allow. Detailed guidance on accommodating the needs of persons with disabilities is available from various sources, including the National Fire Protection Association and the United States located National Organization on Disability [302, 303].

In certain occupancies occupants may be slow in their responses and should be accommodated accordingly. Occupants are likely to be sleeping in residential and healthcare occupancies. Occupants may be intoxicated in night clubs and bars. In many occupancies, such as places of assembly, transportation terminals and hotels, occupants may not move until directed by staff.

13.9 Procedural Assistance

To the extent that occupants are unlikely to move to safer locations without assistance, help should be provided. Help can be provided by organized and trained building emergency response teams. In addition to an occupant movement manager, roles for building emergency response teams can include responding fire fighters, security personnel, floor wardens or monitors, persons charged with searching areas to ensure that everyone has left (assuming that they have been instructed to move to another location), persons assigned to help persons with disabilities, and elevator monitors who prevent occupants from using elevators or, when the building is equipped with elevators that can be safely used, directing and reassuring occupants that they can safety wait for an elevator car.

In buildings where one-way public address systems or an automated alarm notification are installed, Instructions about who should locate where can be provided to building occupants. (See Chap. 12, Enhancing Human Response to Emergency Notification and Messaging) Building emergency response teams may still be required to ensure that broadcasted messages are understood and to reinforce the feasibility of the recommended actions when occupants have doubts. Building occupants may be unfamiliar with the layout of the building, especially egress routes. This is the likely case not only for visitors, but also for occupants who have not participated in fire drills designed to familiarize them with alternative means to egress from the building. Code requirements for certain occupancies, such as apartment buildings, typically omit requirements for communicating instructions during emergencies. In this situation, appropriate education for building occupants is required. As indicated by using the decision model, in well-compartmented buildings, it is often safer to defend occupants where they are already located. Because occupants may choose to move without any clear knowledge of the fire hazards they may encounter, fire protection engineers might consider recommending the installation of emergency communications systems even in the absence of code requirements. Along with public address, addressable telephone systems and intercoms, occupant movement managers can potentially leverage social media capabilities to provide text messages to building occupants.

As a generalization, occupants assessed as most in danger from fire hazards should be assigned the highest priority in receiving information that recommends the actions that they should take. Occupants in relatively less danger may be requested to wait for further instructions that depend on how the fire emergency is developing. Regardless, it is important to communicate with building occupants who are not in immediate danger. They are likely to make decisions in the absence of advice from occupant movement managers, potentially leading them to make mistakes could that endanger themselves and rescuers by moving to more vulnerable locations or interfering with the movement of occupants in greater jeopardy.

A building emergency response team may also be responsible for assisting persons who are not able or who are unlikely to evacuate without assistance. While some buildings maintain registers of non-transient occupants who are likely to require assistance, these cannot be relied on to be complete. Apart from the problem of maintaining up-to-date registers, registers are likely to omit visitors who have limitations, persons who have temporary disabilities and persons who have not voluntarily included themselves in the register.

13.10 Using the Operational Model to Adapt the Plan Depending on How the Emergency Develops

In addition to planning for fire emergencies, the decision model for the operational management of occupant movements can be used to change occupant movement strategies *during* a fire emergency. When everything goes according to a well-conceived plan, the manager will recommend timely and appropriate responses that will safeguard all the building occupants during the entire emergency. But this cannot be guaranteed. There are inherent uncertainties in the way that fire emergencies evolve—and the more complex the situation, the more likely that everything will not happen as planned. "A single solution may not be appropriate for complex facilities. These facilities may benefit from adaptive or event-based strategies. With adaptive or even-based strategies, conditions dictate the specific actions, and egress strategy. During an event, the situation is assessed and a strategy is selected based on that assessment" [290].

13.10.1 Examples of Circumstances that Can Disrupt Even Well-Conceived Plans

13.10.1.1 The Fire is More Severe than Anticipated

Over time, fuel loads can exceed those originally anticipated. Intentionally instigated explosions and fires can result in severe fire conditions.

13.10.1.2 Failures of Fire Protection Features

Fire protection features may fail to operate as intended. Fire protection engineers and fire code enforcers can stress the importance of assessing the condition and maintenance of fire protection features needed to ensure that occupants will remain safe during the duration of a fire. Faults can arise from design, construction and maintenance errors. Separations may have penetrations and doors may not close and latch allowing smoke to spread to other areas in the building. It is not uncommon for separations of protected vertical openings and adjacent horizontal spaces to be violated by contractors when installing conduits after building are occupied. Alarm and communication systems can fail, delaying notification of a fire hazard and preventing alerts and instructions from being transmitted. Fire sprinklers are often considered to be the most effective and reliable fire protection features. However, occupant movement managers should understand that failures in sprinkler systems are unlikely but possible due to errors in building design, construction and maintenance [304]. Sprinklers require testing and building management and fire department personnel can inspect premises to ensure that sprinkler heads remain unobstructed. Further, building management should always know when fire protection features are out of service, and be prepared to alter the plan for safeguarding occupants. Whether systems are not reliably maintained or are out of service, this will often change the answer to the question about whether occupants will remain safe in stationary locations and when moving to new locations.

13.10.1.3 Organizational Failures

Persons charged with key roles during a fire emergency may not remember to perform items in a protocol (as noted earlier, nursing staff may fail to close patient room doors), may be absent, or replacements may not have been appointed.

13.10.1.4 Inaccurate Assessments or Unknown Capabilities of Building Occupants

In many occupancies, the capabilities of building occupants are well understood. Families know the limitations of their members. Health care facilities generally assume low levels of capability among their patients. But in other occupancies, the presence of persons with limitations is difficult to assess. Hotels are unlikely have a clear understanding of the capabilities of their guests. Office buildings may maintain a register of persons who are likely to need assistance, but even the most comprehensive registers will have omissions due to temporary disabilities and the presence of guests. Moreover, persons may suffer injuries during the incident that can change their abilities to move to safer locations without assistance.

13.10.1.5 Complex Systems Interact with their Environments in Ways that Cannot Always be Anticipated

Disastrous outcomes to emergencies usually result from a convergence of events that are arguably impossible to anticipate. Components of complex systems can interact with each other and with environmental conditions in ways that are unexpected [305, 306].

Because of the above sources of uncertainty, emergencies may not play out the way that the planning and design scenarios describe events. Good situation awareness (see Sect. 13.10.4) improves the ability of the occupant movement manager to adapt to unforeseen (perhaps unforeseeable) events, even when a well-considered set of scenarios is used to plan for fire emergencies.

13.10.2 Backup Strategies

Backup strategies may be required depending on how well strategies work relative to the extent of fire hazards. One form of planning backup strategies is described by Bukowski and Tubbs [290] as "a scalable approach that escalates from partial evacuation and protect-in-place to simultaneous full building evacuation as necessary. In certain residential buildings, means of escape (i.e., ways to leave the building that do not qualify as "exits") are provided in the event that the primary route is unavailable. As described in the model for the operational management of occupant movement, problems may force some occupants to take refuge while awaiting rescue. The successful use of backup strategies depends on maintaining good situation awareness. Occupant movement managers can incorporate backup strategies as needed into their planning, thereby speeding their use depending as a fire develops.

13.10.3 Using the Model During Fire Emergencies

When applying the model during a fire emergency, the situation is periodically assessed by reevaluating the first decision: "Which groups of occupants are safe where they are already located?" and "Are there reliable means to move the group to a safer location?" The required information remains the same: the locations of occupants, the projected locations of hazards, the degree that occupants are likely to remain separated from the hazard, and the limitations of occupants and the availability of assistance to compensate for those limitations. To answer these questions, occupant movement managers need to maintain some degree of situation awareness. The process of monitoring the situation is shown in the feedback loops in Fig. 13.5.

13.10.4 Definition of and Importance of Situation Awareness while Managing Occupant Movement During Fire Emergencies

For all the decisions in the model, a dynamic assessment of both the extent of the hazard and the locations and capabilities of occupants is important. "Appropriate systems and methods are needed for decision makers to quickly obtain credible information about the event and to empower decision makers with the appropriate authority to make egress decisions based on that information" [290]. Situation awareness [307] has been defined by Groner in emergency settings as "the degree that occupants responding to an emergency (1) are aware of the situation in which they find themselves, (2) understand the meaning of the situation as it affects their abilities to pursue goals, and (3) accurately anticipate how the situation is likely to change as time passes" [308].

The last level of situation awareness, anticipating how an emergency is likely to change, is particularly problematic. Like most people lacking an understanding of fire dynamics, persons managing an evacuation may have an inaccurate understanding of how quickly fire hazards can develop. For this reason, occupant movement managers should have a basic understanding of fire dynamics such as the exponential growth of fire, flashover, and the tendency of fire and smoke to move upwards where unimpeded. In maintaining good situation awareness, occupant movement managers should also have a basic understanding of how people tend to behave during fires (see Chap. 4). Research supports the view that people can generally be relied on to follow instructions from a trusted source provided that those instructions seem consistent with their individualized situation awareness.

Establishing good situation awareness involves collecting and making sense of the types of information indicated in the occupant movement models discussed in this chapter. Information can be obtained from building occupants, emergency response team member, and interfaces with various building systems.

In the context of larger buildings or transportation systems using warning or messaging systems, often involving a two-stage alarm system, the time to provision of a warning to general occupants is often heavily dependent on the staff response to the pre-alarm. When a detection event activates a pre-alarm to security, staff enter their own sub-routine of pre-response delays and responses before issuing a general warning to other affected occupants. After an initial delay the first response of staff is often to investigate the source. They then report their findings up the management chain, sometimes resulting in a sequence of further investigation and reporting until an incident is either resolved without alerting other occupants or is considered sufficiently serious to issue an evacuation warning to affected occupants. In some situations, occupants becoming aware of a fire during this period may be actually be discouraged from evacuating. Pre-warning delays resulting from these processes have led to many deaths during fire incidents. It is therefore important to design staff emergency response systems and procedures to minimize such delays, train staff to recognize potentially serious developing emergencies as early as possible and to take appropriate action, employing a dynamic risk assessment process. Where a stay-put or phased evacuation plan is in place (for example in high rise buildings), then it is important to keep all occupants informed of a developing situation, so that all occupants (including those on standby) are able to evacuate if and when they choose to do so.

Personal communications are essential means for obtaining good situation awareness by persons managing occupant movement. In addition to relaying recommended strategies for groups of building occupants, interpersonal communications are used to: (1) assess the extent of fire hazards; (2) assess the locations and capabilities of building occupants; and, (3) determine the availability of assistance for occupants who may be unable to take recommended protective actions on their own. Communications devices are essential in larger complex buildings. Security personnel and fire fighters typically have radios. Building emergency team members (e.g., floor wardens, fire safety directors) typically use intercoms and perhaps landline and mobile phones to communicate important information. Calls to public safety answering points (PSAPs), more typically called emergency dispatch centers, can provide valuable situation awareness information, although without training and planning, failures to relay this information to occupant movement managers have resulted in loss of life [255].

The most important hardware interface for obtaining situation awareness information is probably a fire alarm panel or alarm annunciator panel. The manager should understand the limitations of the annunciator, for example, whether an activation indicates a zone or a specific addressable detection device. Annunciators may also provide valuable information about the status of elevator cars—their locations and whether they are available for service or can only be used by fire fighters. Closed-circuit television systems (CCTV) have the potential to provide valuable information to occupant movement managers. Some large buildings have CCTV systems with cameras in egress routes, including exit stairs and building exteriors, albeit for the purposes of security and not the emergency movement of occupants. These can provide useful information about smoke conditions and the flow of persons using egress routes. In most instances, CCTV systems are monitored from security stations that are separate from where fire alarm systems are monitored and fire departments stage their efforts.

13.11 Guidance on Implementing the Models

13.11.1 Informational Inputs Needs to be Acquired from a Variety of Sources

Any single person is very unlikely to have all of the informational required as inputs to the decision models described in this chapter. Meetings that include all of persons who can provide information are encouraged. It can be important that people in various design and operational roles understand each other's needs for information. Understanding the goals and objectives of persons in different roles encourages acceptance to the strategic objectives of people moving planning and a willingness to collect and transfer needed information in the design, planning and real-time emergency phases.

During the design phase, the design team comprised of fire protection engineers, other design engineers, architects and property developers, should acquire information from persons knowledgeable about the building once it is occupied. These persons include building managers and facility engineers acquainted with similar buildings during their operational phase.

Once the building is occupied, building managers, emergency team members, tenants, and responding fire fighters should have detailed information about the building and its fire protection features.

During an emergency, it is important that building management and responding fire fighters and other public safety personnel are pursuing the same strategic and tactical objectives. When people in different roles are not in agreement, misleading and contradictory information may be passed on to building occupants, leading to their unnecessary confusion and endangerment. It is important to note that an occupant movement plan should be not a lengthy document that gathers dust on a shelf. The models described in this chapter are simple enough to be easily understood and recalled by everyone with an occupant movement function. The "action" plan derived from the models should be distributed so that the resulting instructions should be simple, straightforward and tailored to specific roles. For example, a floor warden can be instructed to reinforce messages broadcast by a public-address system by explaining why occupants should leave the building, relocated to another part of the building, or remain in their present locations.

Finally, an occupant movement plan is only a part of a comprehensive fire emergency plan. As only one example, a good plan includes training for a building emergency response team and occupants regarding the appropriate actions when discovering a suspected fire, such as activating the alarm system and notifying the fire department, security services and other responsible parties.

13.11.2 A Checklist for Data Inputs to the Decisions in the Models

The following sections include a list of informational inputs that are related to the decision processes described in the models. The list is intended as general guidance, and there may be omissions that should nonetheless be taken into account when prescribing occupant movement strategies. More detailed information is available in this and other chapters of this guide, as well as from other sources [309–311].

13.11.2.1 Building Characteristics and Fire Protection Features
- Areas of refuge and rescue-assistance
- The building layout including floor plans and means of egress

13.11 Guidance on Implementing the Models

- Fire protection features that mitigate the development of fire, including fire extinguishers (assuming that persons are trained in their use, automatic suppression (including sprinklers and chemical extinguishing systems for cooking surfaces and computer and data centers), and smoke control systems
- Fire protection features that separate occupants from hazards while occupants are stationary, including fire- and smoke-rated assemblies, other non-rated barriers, dampers and other means that limit the spread of smoke through HVAC systems, and horizontal means of egress, exits, and non-code conforming means of egress,
- Fire protection features that separate occupants from hazards while moving to safer location using means of egress, including horizontal exits, exit corridors and stairs, non-code conforming means of egress, and means of egress discharge that lead occupants to safe location outside the building, smoke-proof exit stairs, protected elevators, and elevators that can only be safely operated during phase II operation by fire fighters and other qualified professionals.
- Fire detection devices and notification appliances, including alarm annunciators, smoke and fire detectors, off-site notification, manual pull-stations, public address systems, backup batteries or generators, alarm signaling appliances, and active and passive signage

13.11.2.2 Maintenance and Inspections

- Regulatory-required inspections of mitigation and alarm features
- Notification during maintenance and retrofits that disable fire protection features
- Inspection following restoration of fire protection features to ensure operability, including penetrations of smoke- and fire-rated barriers
- Exit drills, including those the exceed regulatory requirement
- Obstructed and code-complying means of egress, including signage, lighting, and backup power supplies

13.11.2.3 Occupant Characteristics

- Locations of occupants at the time of the scenario under consideration
- Mobility-related disabilities, including permanent, temporary and hidden limitations
- Sensory-related disabilities, including sight and hearing
- Cognitive limitations, including dementia, developmental disabilities, and the likelihood of intoxication at the time of the scenario under consideration
- Sleeping occupants at the time of the scenario under consideration
- Occupants with assignments that may delay their departure from a particular location

13.11.2.4 Availability of Assistance

Staffing, including extent of training and number of staff available at the time of the scenario under consideration.

- Emergency team members, including fire safety directors, fire/floor wardens, elevator monitors, and persons assigned to assist occupants with disabilities.
- Arrival times and available resources for responding fire fighters and other public safety personnel
- Evacuation chairs (controlled decent devices) and persons trained in their operation.
- Alarm notification devices, including non-informational signals (e.g., alarm bells) and public-address systems
- Two-way communications, including intercoms, cell phones and radios.
- Active and passive signage

Addendum: Glossary of Terms

Alarm A signal or message from a person or device indicating the existence of an emergency or other situation that may require action

Area of refuge An area that is a space located in a path of travel that is protected from the effects of fire, either by means of separation from other spaces in the same building or by virtue of location or other protective measures, thereby permitting a delay in egress travel

Asphyxiant Toxicant that causes reduced oxygen reaching body tissues, which can result in central nervous system depression or cardiovascular effects, potentially resulting in Loss of consciousness and death

Audible notification appliance A notification appliance that alerts by the sense of hearing.

Available safe egress time (ASET) The calculated time interval between the time of ignition and the time at which conditions become such that the occupant is estimated to be incapacitated, i.e. unable to take effective action to escape to a safe refuge or place of safety (time available for an individual occupant to escape or move to a safe location).

Built environment Building or other occupiable structure

Ct product The concentration–time product ((μL/L) min) obtained by the integration of the area under a plot of the concentration of a toxic gas as a function of time

Defend in place Life safety strategy in which occupants are encouraged to remain in their current location rather than to attempt escape during a fire; or, the operational response in which the action is to relocate the affected occupants to a safe place within the structure during an emergency.

Evacuation plan A plan specifying safe and effective methods for the movement of people from locations threatened by fire.

Evacuation time Time interval between the time of a warning of fire being transmitted to the occupants and the time at which an occupant of a specified part of a building or all of the building is able to enter a place of safety.

Exit A designated point of departure from a building, or a doorway or other suitable opening giving access toward a place of relative safety, or that portion of a means of egress that is separated from all other spaces of a building or structure by construction or equipment as required to provide a protected way of travel.

Exposure dose Measure of the maximum amount of a toxic gas that is available for inhalation, calculated by integration of the area under a plot of the concentration of a gas as a function of time

Exposure time Length of time for which people, animals or test specimens are exposed under specified conditions

Fire effluent Totality of gases and aerosols, including suspended particles, created by combustion or pyrolysis in a fire. Note: In this document, this is synonymous with smoke.

Fire scenario Qualitative description of the course of a fire with respect to time, identifying key events that characterize the studied fire and differentiate it from other possible fires and which may include the ignition and fire growth processes, the fully developed fire stage, the fire decay stage, and the environment and systems that impact on the course of the fire.

Heat stress Conditions caused by exposure to elevated or reduced temperature, radiant heat flux, or a combination of these factors

Hyperventilation Rate and/or depth of breathing which is greater than normal

Incapacitation State of physical inability to accomplish a specific task, such as accomplishing escape from a fire.

Irritant Gas or aerosol that stimulates nerve receptors in the eyes, nose, mouth, throat and respiratory tract, causing varying degrees of discomfort and pain with the initiation of numerous physiological defense responses. Note: Physiological defense responses include reflex eye closure, tear production, coughing, and bronchoconstriction.

Means of egress A continuous and unobstructed way of travel from any point in a building or structure to a public way.

Movement time Time needed for an occupant to move to and through a means of egress or means of escape passing to a place of safety

© Society of Fire Protection Engineers 2019
SFPE Society of Fire Protection Engineers, *SFPE Guide to Human Behavior in Fire*, https://doi.org/10.1007/978-3-319-94697-9

Occupant load The total number of persons that might occupy a building or portion thereof at any one time.

Place of safety Location which is assumed free from danger and from which it is possible to move freely without threat from a fire.

Pre-evacuation time Interval between the time at which a warning of fire is given and the time at which the first move is made towards an exit

Required safe egress time (RSET) The calculated time period required for an individual occupant to travel from their location at the time of ignition to a safe refuge or place of safety (time required for an individual occupant to escape or move to a safe location).

Safe location (place of safety) A location remote or separated from the effects of a fire inside or outside the building depending upon the evacuation strategy so that such effects no longer pose a threat,

Safety factor A factor applied to a predicted value to ensure that a sufficient safety margin is maintained.

Safety margin The difference between a predicted value and the actual value where a fault condition is expected.

Situation awareness The perception of the elements in the environment within a volume of time and space, the comprehension of their meaning, and the projection of their status in the near future.

Smoke Totality of gases and aerosols, including suspended particles, created by combustion or pyrolysis in a fire. Note: In this document, this is synonymous with fire effluent. However, in some of the references, "smoke" refers only to the visible part of "smoke."

Tenability Environmental conditions in which smoke and heat are at sufficiently low levels that the impact on occupants is not incapacitating or life-threatening.

Travel distance Distance that is necessary for a person to travel from any point within a built environment to the nearest exit or safe location, taking into account the layout of walls, partitions, stairs and fittings

Travel time The time needed, once movement towards an exit has begun, for an occupant of a specified part of a building to reach a safe location.

Yield Mass of a combustion product generated during combustion or pyrolysis divided by the mass loss of the test specimen or, in a fire, the combustible item of interest

References

1. Society of Fire Protection Engineers, Engineering Guide: Human Behavior, Bethesda, MD: Society of Fire Protection Engineers, 2003.
2. Society of Fire Protection Engineers, The SFPE Handbook of Fire Protection Engineering, 5 ed., M. J. Hurley, Ed., Gaithersburg, MD: Society of Fire Protection Engineers, 2016.
3. Society of Fire Protection Engineers, Performance-Based Fire Protection, 2 ed., Quincy, MA: National Fire Protection Association, 2007.
4. R. F. Fahy and G. Proulx, "Toward Creating a Database on Delay Times to Start Evacuation and Walking Speeds for use in Evacuation Modeling," in *Second International Symposium on Human Behavior in Fire*, 2001.
5. F. I. Stahl, J. J. Crosson and S. T. Margulis, "Time-based Capabilities of Occupants to Escape Fire in Public Buildings: A Review of Code Provision and Technical Literature," National Bureau of Standards, Gaithersburg, MD, 1982.
6. R. Gann, "Final Report on the Collapse of the World Trade Center Towers," NIST, Gaithersburg, MD, 2005.
7. International Standards Organization, "Fire Safety- Vocabulary".
8. J. A. Milke and T. C. Caro, "A Survey of Occupant Load Factors in Contemporary Office Buildings," *Journal of Fire Protection Engineering*, vol. 8, no. 4, pp. 169–182, November 1996.
9. "Special Issue: World Trade Center Bombing," *Fire Engineering*, vol. 96, no. 12, December 1993.
10. "World Trade Center Response: An Event Both Technical and Personal," *WNYF*, vol. 54, no. 3, 1993.
11. G. Proulx, "Occupant Behaviour and Evacuation," National Research Council Canada, Ottawa, Canada, 2001.
12. D. Nilsson and A. Johansson, "Social influence during the initial phase of a fire evacuation—Analysis of evacuation experiments in a cinema theatre," *Fire Safety Journal*, vol. 44, no. 1, pp. 71–79, 2009.
13. B. Latané and J. M. Darley, "Group Inhibition of Bystander Intervention in Emergencies," *Journal of Personality and Social Psychology*, vol. 10, no. 3, pp. 215–221, 1968.
14. R. Lovreglio, E. Ronchi and D. Borri, "The Validation of Evacuation Simulation Models through the Analysis of Behavioural Uncertainty," *Reliability Engineering and System Safety*, vol. 46, no. 37, pp. 166–174, 2014.
15. R. Lovregilo, A. Fonzone and L. Dell'Olio, "The Role of Herding Behaviour in Exit Choice during Evacuation," *Procedia - Social and Behavioral Sciences*, vol. 160, pp. 390–399, 2014.
16. J. Sime, "Movement Toward the Familiar: Person and Place Affiliation in a Fire Entrapment Setting," *Environment and Behavior*, vol. 17, no. 6, pp. 697–724, November 1985.
17. T. McClintock, T. Shields, A. Reinhardt-Rutland and J. Leslie, "A Behavioural Solution to the Learned Irrelevance of Emergency Exit Signage," in *Proceedings of the Second International Conference of Human Behaviour in Fires*, Boston, MA, 2001.
18. D. Nilsson, H. Frantzich and W. L. Saunders, "Influencing Exit Choice in the Event of a Fire Evacuation," in *Proceedings of the Ninth International Symposium on Fire Safety Science*, Karlsruhe, Germany, 2008.
19. D. Bruck and P. Brennan, "Recognition of Fire Cues During Sleep," in *Second International Symposium on Human Behavior in Fire*, London, 2001.
20. M. Ball and D. Bruck, "The Effect of Alcohol upon Response to Fire Alarm Signals in Sleeping Young Adults," in *The Third International Symposium on Human Behaviour in Fire*, Belfast, Northern Ireland, 2004.
21. K. E. Boyce, T. J. Shields and W. H. Silcock, "Toward the Characterization of Building Occupancies for Fire Safety Engineering: Capabilites of Disabled People Moving Horizontally and on an Incline," *Fire Techology*, vol. 35, no. 1, pp. 51–67, Feburary 1999.
22. K. E. Boyce, T. J. Shields and G. W. Silcock, "Toward the Characterization of Building Occupancies for Fire Safety Engineering: Capability of People with Disabilities to Read and Locate Exit Signs," *Fire Technology*, vol. 35, no. 1, pp. 79–86, 1999.
23. K. E. Boyce, T. J. Shields and W. H. Silcock, "Toward the Characterization of Building Occupancies for Fire Safety Engineering: Capability of Disabled People to Negotiate Doors," *Fire Techonology*, vol. 35, no. 1, pp. 68–78, February 1999.
24. T. J. Shields, B. Smyth, K. E. Boyce and G. W. H. Silcock, "Towards the Prediction of Evacuation Behaviours for People with Learning Difficulties," *Facilities*, vol. 17, no. 9/10, pp. 336–344, 1999.
25. J. D. Sime, "The Outcome of Escape Behaviour in the Summerland Fire: Panic or Affiliation?," in *Proceedings of the International Conference on Building Use and Safety Technology*, Los Angeles, CA, 1985.
26. W. Feinberg, "Primary Group Size and Fatality Risk in a Fire Disaster," in *Second International Symposium on Human Behaviour in Fire*, London, 2001.
27. D. A. Purser and M. Bensilum, "Quantification of Behaviour for Engineering Design Standards and Escape Time Calculations," in *Proceedings of the First International Symposium on Human Behaviour in Fire*, Belfast, U.K., 1998.
28. J. Creak, "'Viewing Distances," *Means of Escape*, 1997.
29. W. Saunders, "Gender Differences in Response to Fires," in *Proceedings of the Second International Symposium on Human Behaviour in Fire*, London, U.K., 2001.
30. J. L. Bryan, "Smoke as a Determinant of Human Behaviour in Fire Situations," National Bureau of Standards, Washington, D.C., 1977.

© Society of Fire Protection Engineers 2019
SFPE Society of Fire Protection Engineers, *SFPE Guide to Human Behavior in Fire*, https://doi.org/10.1007/978-3-319-94697-9

References

31. P. G. Wood, *Fire Research Note 953,* Fire Research Station, 1972.
32. J. L. Bryan and J. A. Milke, "The Determination of Behavior Response Patterns in Fire Situations, Project People II," National Bureau of Standards, Washington, D.C., 1980.
33. J. L. Bryan, "Implications for Codes and Behavior Models from the Analysis of Behavior Response Patterns in Fire Situations as Selected from the Project People and Project People II Study Programs," National Bureau of Standards, Washington, D.C., 1983.
34. A. Ozkaya, "A Qualitative Approach to Children of Developing Countries from Human Behaviour in Fire Aspect," in *Proceeding of the Second International Symposium on Human Behaviour in Fire*, London, U.K., 2001.
35. M. Almejmaj, B. J. Meacham and J. Skorinko, "The Effects of Cultural Differences between the West and Saudi Arabia on Emergency Evacuation - Clothing Effects on Walking Speed," *Fire and Materials,* vol. 39, no. 4, pp. 353–370, 21 January 2014.
36. E. R. Galea, M. Sauter, S. J. Deere and L. Filippidis, "Investigating the Impact of Culture on Evacuation Response Behaviour," in *Proceedings of the Twelfth International Fire Science and Engineering Conference*, London, U.K., 2010.
37. E. R. Galea, M. Sauter, S. Deere and L. Filippidis, "Investigating the Impact of Culture on Evacuation Behavior - A Turkish Data-Set," *Fire Safety Science,* pp. 709–722, 2011.
38. S. Kose, "Emergence of Aged Populace: Who is at Higher Risk in Fires?," in *Proceedings of the First International Symposium*, Ulster, 1998.
39. D. A. Purser, "Toxicity Assessment of Combustion Products," in *SPFE Handbook of Fire Protection Engineering*, 3 ed., Quincy, MA, National Fire Protection Association, 2002, pp. 2-83 - 2-171.
40. R. G. Gann, J. D. Averill, K. M. Bulter, W. W. Jones, G. W. Mulholland, J. L. Neviaser, T. J. Ohlemiller, R. D. Peacock, P. A. Reneke and J. R. Hall Jr., "International Study of the Sublethal Effects of Fire Smoke on Survivability and Health (SEFS): Phase 1 Final Report," National Institue of Standards and Tecnhonolgy, Gaithersburg, 2001.
41. International Standards Organization, "Life-threatening Components of Fire - Guidelines for the Estimation of Time Available for Escape using Fire Data," Geneva, Switzerland, 2002.
42. G. Proulx and J. D. Sime, "To Prevent 'Panic' in an Underground Emergency: Why Not Tell People the Truth?," in *Proceedings of the third International Symposium of Fire Safety Science*, London, U.K., 1991.
43. G. Proulx and R. F. Fahy, "The Time Delay to Start Evacuation: Review of Five Case Studies," in *Proceedings of the Fifth International Symposium of Fire Safety Science*, London, U.K., 1997.
44. E. Cable, "An Analysis of Delay in Staff Response to Fire Alarm Signals in Health Care Occupancies," Worcester Polytechnic Institute, Worcester, MA, 1993.
45. J. H. Sorensen and B. V. Sorensen, "Community Processes: Warning and Evacuation," in *Handbook of Disaster Research*, 1 ed., H. Rodriguez, E. L. Quarantelli and R. Dynes, Eds., New York, NY, Springer-Verlag, 2007, pp. 183–199.
46. D. S. Mileti and L. Peek-Gottschlich, "Hazards and Sustainable Development in the United States," *Risk Management,* vol. 3, no. 1, pp. 61–70, 2001.
47. K. J. Tierney, M. K. Lindell and R. W. Perry, Facing the Unexpected: Disaster Preparedness and Response in the United States, Washington, D.C.: Joseph Henry Press, 2001.
48. D. S. Mileti, T. E. Drabek and E. J. Haas, "Human Systems in Extreme Environments: A Sociological Perspective," Institute of Behavioral Science, University of Colorado, Boulder, CO, 1975.
49. M. K. Lindell and R. W. Perry, Communicating Environmental Risk in Multiethnic Communities, Thousand Oaks, CA: Sage Publications, 2004.
50. D. S. Mileti and J. H. Sorensen, "Communications of Emergency Public Warnings: A Social Science Perspective and State-of-the-art Assessment," National Laboratory, U.S. Department of Energy, Oak Ridge, TN, 1990.
51. J. D. Averill, D. S. Mileti, R. D. Peacock, E. D. Kuligowski, N. Groner, G. Proulx, P. A. Reneke and H. E. Nelson, "Occupant Behavior, Egress, and Emergency Communication. Federal Building and Fire Safety Investigation of the World Trade Center Disaster," National Institute of Standards and Technology, 2005.
52. J. Lynch, "Nocturnal Olfactory Response to Smoke Odor," in *Human Behaviour in Fire: Proceedings of the First International Symposium*, Univeristy of Ulster, Belfast, U.K., 1998.
53. T. Jin, "Studies on Human Behavior and Tenability in Fire Smoke," in *The Fifth International Symposium on Fire Safety Science*, 1997.
54. D. Bruck, "Non-awakening in Children in Response to a Smoke Detector Alarm," *Fire Safety Journal,* vol. 32, pp. 369–376, 1999.
55. E. H. Nober, H. Peirce, A. D. Well, C. C. Johnson and C. Clifton, "Waking Effectiveness of Household Smoke and Fire Detection Devices," National Institute of Standards and Technology, Gaithersburg, MD, 1980.
56. National Fire Protection Association, NFPA 72® National Fire Alarm and Signaling Code®, Quincy, MA, 2016.
57. E. H. Nober, A. Well and S. Moss, "Alarms for the Hearing-Impaired," *Fire Prevention,* no. 233, pp. 28–31, October 1990.
58. E. M. Ashley, "Waking Effectiveness of Emergency Alerting Devices for the Hearing Able, Hard of Hearing Dean Populations," College Park, 2007.
59. S. M. V. Gwynne, "Optimizing Fire Alarm Notification for High Risk Groups: Notification Effectiveness for Large Groups," prepared for The Fire Protection Research Foundation, Quincy, MA, 2007.
60. M. Ball and D. Bruck, "The Salience of Fire Alarm Signals for Sleeping Individuals: A Novel Approach to Signal Design," in *The Third International Symposium on Human Behavior in Fire*, Belfast, Northern Ireland, 2004.
61. R. Timmons, "Interoperability: Stop Blaming the Radio," *Homeland Security Affairs,* vol. III, no. 1, pp. 1–17, February 2007.
62. R. P. Timmons, "Sensory Overload as a Factor in Crisis Decision-making and Communications by Emergency First Responders," The University of Texas at Dallas, Dallas, TX, 2009.
63. E. Cable, "Cry Wolf Syndrome: Radical Changes Solve the False Alarm Problem," in *The Fourth National Symposium & Trade Exhibition on Health Care Safety and the Environment*, 1994.
64. G. Proulx, "Occupant Response to Fire Alarm Signals," in *National Fire Alarm Code Handbook*, Quincy, MA, National Fire Protection Association, 1999.
65. G. Proulx, C. Laroche, F. Jaspers-Fayer and R. Lavallée, "Fire Alarm Signal Recognition," National Research Council Canada, 2001.
66. R. Chandler, Emergency Notification, Santa Barbara, CA: Praeger, 2012.
67. R. F. Fahy and G. Proulx, "Human Behavior in the World Trade Center Evacuation," in *Fire Safety Science - Fifth International Symposium.*

References

68. G. Proulx, J. Pineau, J. C. Latour and L. Stewart, "Study of the Occupants' Behaviour during the 2 Forest Laneway Fire in North York, Ontario, January 6, 1995," National Research Council of Canada, Ottawa, Canada, 1995.

69. G. Ramachandran, "Informative Fire Warning Systems," *Fire Technology,* vol. 27, no. 1, pp. 66–81, February 1991.

70. S. Breznitz, Cry Wolf: The Psychology of False Alarms, Hillsdale, New Jersey London: Lawrence Erlbaum Associates, 1984.

71. D. Breen, "Do Smoke Detection Systems Work in College Dormitories?," Society of Fire Protection Engineers, Bethesda, MD, 1984.

72. I. Janis and L. Mann, Decision Making: A Psychological Analysis of Conflict, Choice, and Commitment, New York, NY: The Free Press, A Division of Macmillan Inc., 1979.

73. P. Wright, "The Harassed Decision Maker: Time Pressures, Distractions, and the Use of Evidence," *Journal of Applied Physchology,* vol. 59, no. 5, pp. 551–561, 1974.

74. T. Engländer and T. Tyszka, "Information Seeking in Open Decision Situations," *Acta Psychologica,* vol. 45, no. 3, pp. 169–176, August 1980.

75. R. Groner, M. Groner and W. F. Bischof, "The Role of Heuristics in Models of Decision," *Advances in Psychology,* vol. 16, pp. 87–108, 1983.

76. H. A. Simon, Models of Man: Social and Rational Mathematical Essays on Rational Human Behavior in a Social Setting, New York, NY: John Wiley and Sons, Inc., 1957.

77. J. Raub, Environmental Health Criteria: Carbon Monoxide, 2 ed., Geneva: World Health Organization, 2004.

78. J. L. McAllister, Interviewee, *Personal Communication.* [Interview]. 30 December 2014.

79. D. A. Purser, P. Grimshaw and K. R. Berrill, "Intoxication by Cyanide in Fires: A Study in Monkeys using Polyacrylonitrile," *Archives of Environmental Health,* vol. 39, no. 6, pp. 394–400, November/December 1984.

80. J. L. Bryan, "An Examination and Analysis of the Dynamics of the Human Behavior in the MGM Grand Hotel Fire," National Fire Protection Assoication, Quincy, MA, 1983.

81. G. Proulx and I. Reid, "Human Behavior Study, Cook County Administration Building Fire, October 17, 2003 - Chicago, IL," National Research Council of Canada, Ottawa, Canada, 2003.

82. W. Grosshandler, N. Bryner, D. Madrzykowski and K. Kuntz, "Report of the Technical Investigation of the Station Nightclub Fire," National Insititute of Standards and Technology, 2005.

83. K. Fridolf, E. Ronchi, D. Nilsson and H. Frantzich, "Movement Speed and Exit Choice in Smoke-filled Rail Tunnels," *Fire Safety Journal,* vol. 59, pp. 8–21, July 2013.

84. D. Nilsson, "Exit Choice in Fire Emergencies - Influencing Choice of Exit with Flashing Lights," Lund University, Lund, Sweden, 2009.

85. H. Frantzich, "Occupant Behaviour and Response Time - Results from Evacuation Experiments," in *Second Symposium on Human Behaviour in Fire,* London U.K., 2001.

86. B. Latané and J. M. Darley, The Unresponsive Bystander: Why Doesn't He Help?, New York, NY: Appleton-Century Crofts, 1970.

87. T. J. Shields, K. Boyce and G. Silcock, "Towards the Characterization of Large Retail Stores," in *Proceedings of the First International Symposium on Human Behavior in Fire,* Univeristy of Ulster, Belfast, U.K., 1998.

88. J. A. Swartz, "Human Behavior in the Beverly Hills Fire," *Fire Journal,* vol. 73, no. 3, pp. 73–74, May 1979.

89. P. Keating, "Human Response during Fire Situations: A Role for Social Engineering," in *Proceedings of Research and Design 85: Architectural Applications of Design and Technology Research,* Los Angeles, CA, 1985.

90. I. Donald and D. Canter, "Behavioral Aspects of the King's Cross Disaster," *Fires and Human Behavior,* pp. 15–30, 1990.

91. Federal Emergency Management Agency, "Establishing a Relationship Between Alcohol and Casualties of Fire," 2003.

92. Federal Emergency Management Agency, "Fraternity and Sorority House Fires," 2002.

93. W. S. Hollis, "Drinking: Its Part in Fire Deaths," *Fire Journal,* vol. 67, no. 3, pp. 10–11, May 1973.

94. W. G. Berl and B. M. Halpin, "Human Fatalities from Unwanted Fires," *Johns Hopkins APL Technical Digest,* vol. 1, no. 2, pp. 129–134, 1980.

95. S. W. Marshall, C. W. Runyan, S. Bangdiwala, M. A. Linzer, J. J. Sacks and J. D. Butts, "Fatal Residential Fires: Who Dies and Who Survives?," *Journal of the American Medical Association,* vol. 279, no. 20, pp. 1633–1637, 1998.

96. D. J. Barillo, B. F. Rush, R. Goode, R. L. Lin, A. Freda and E. J. Anderson, "Is Ethanol the Unknown Toxin in Smoke Inhalation Injury?," *The American Surgeon,* vol. 52, no. 12, pp. 641–645, December 1986.

97. B. Levine, Principles of Forensic Toxicology, Second Edition, Washington, D.C.: American Association for Clinical Chemistry, 2006.

98. D. Bruck, "The Who, What, and Why of Waking to Fire Alarms: A Review," *Fire Safety Journal,* vol. 36, pp. 623–639, 2001.

99. A. Hasofer and I. T. Thomas, "Sound Intensity Required for Waking Up," *Fire Safety Journal,* vol. 42, no. 4, pp. 265–270, 2007.

100. D. S. Mitchell, S. C. Packham and W. E. Fitgerald, "Effects of Ethanol and Carbon Monoxide on Two Measures of Behavioral Incapacitation of Rats," *Proceedings of the Western Pharmacology Society,* vol. 21, pp. 427–431, 1978.

101. E. D. Kuligowski, "Terror Defeated: Occupant Sensemaking, Decision-making and Protective Action in the 2001 World Trade Center Disaster," University of Colorado at Boulder, Boulder, CO, 2011.

102. J. D. Sime, "The Concept of 'Panic'," in *Fires and Human Behaviour,* D. Canter, Ed., Chichester, Wiley, 1980, pp. 63–81.

103. J. P. Keating, "The Myth of Panic," *Fire Journal,* vol. 76, no. 3, pp. 57–61, May 1982.

104. E. L. Quarantelli, "Panic Behavior: Some Empirical Observations," in *Human Reponse in Tall Buildings,* Stroudsburg, Pennsylvania, Dowden, Hutchinson and Ross, Inc., 1977, pp. 335–350.

105. K. Okabe, "A Study on the Socio-Phychological Effect of a False Warning of the Tokai Earthquake in Japan," in *A Paper Presented at the Tenth World Congress of Sociology,* Mexico City, Mexico, 1982.

106. P. Brennan, "Smoke Gets in Your Eyes: The Effect of Cue Perception on Behaviour in Smoke," in *International Conference on Fire Science and Engineering First Proceedings,* London, England, 1995.

107. E. D. Kuligowski, R. D. Peacock, P. A. Reneke, E. Wiess, C. R. Hagwood, K. J. Overholt, R. P. Elkin, J. D. Averill, E. Ronchi, B. L. Hoskins and M. Spearpoint, "Movement on Stairs During Building Evancations," National Institute of Standards and Technology, Gaithersburg, MD, 2015.

108. D. S. Mileti and J. D. Darlington, "The Role of Searching in Shaping Reactions to Earthquake Risk Information," *Social Problems,* vol. 44, no. 1, pp. 89–103, 1997.

109. T. E. Drabek and J. S. Stephenson III, "When Disaster Strikes," *Journal of Applied Social Psychology,* vol. 1, no. 2, pp. 187–203, June 1971.

110. C. E. Fritz and E. S. Marks, "The NORC Studies of Human Behavior in Disaster," *Journal of Social Issues,* vol. 10, no. 3, pp. 26–41, 1954.

111. D. S. Mileti and C. Fitzpatrick, "The Causal Sequence of Risk Communication in the Parkfield Earthquake Prediction Experiment," *Risk Analysis,* vol. 12, no. 3, pp. 393–400, 1992.
112. D. S. Mileti, "Natural Hazard Warning Systems in the United States: A Research Assessment," Institute of Behavioral Science, The University of Colorado, Boulder, CO, 1975.
113. R. W. Perry, M. K. Lindell and M. R. Greene, Evacuation Planning in Emergency Management, Lexington, MA: Lexington Books, 1981.
114. R. W. Perry and M. R. Green, "The Role of Ethnicity in the Emergency Decision-making Process," *Sociological Inquiry,* vol. 52, no. 4, pp. 306–334, 1982.
115. M. K. Lindell, R. W. Perry and M. R. Greene, "Individual Response to Emergency Preparedness Planning near Mt. St. Helens," *Disaster Management,* no. 3, pp. 5–11, January/March 1983.
116. C. I. Hovland, I. L. Janis and H. H. Kelley, Communication and Persuasion; Psychological Studies of Opinion Change, New Haven : Yale University Press, 1953.
117. A. A. Peguero, "Latino Disaster Vulnerability: The Dissemination of Hurricane Mitigation Information Among Florida's Homeowners," *Hispanic Journal of Behavioral Sciences,* vol. 28, pp. 5–22, 2006.
118. C. W. Trumbo and K. A. McComas, "The Function of Credibility in Information Processing for Risk Perception," *Risk Analysis,* vol. 23, no. 2, pp. 343–353, April 2003.
119. J. Flynn, P. Slovic and C. K. Mertz, "The Nevada Initiative: A Risk Communication Fiasco," *Risk Analysis,* vol. 13, no. 5, pp. 497–508, October 1993.
120. F. N. Burkhart, "Media, Emergency Warnings, and Citizen Response," 2011.
121. R. W. Perry and L. Nelson, "Ethnicity and Hazard Information Dissemination," *Environmental Management,* vol. 15, no. 4, pp. 581–587, 1991.
122. T. E. Drabek and K. S. Boggs, "Families in Disaster: Reactions and Relatives," *Journal of Marriage and the Family,* vol. 30, no. 3, pp. 443–451, August 1968.
123. M. R. Greene, R. W. Perry and M. K. Lindell, "The March 1980 Eruptions of Mt. St. Helens: Citizen Perceptions of Volcano Threat," *Disasters,* vol. 5, no. 1, pp. 49–66, 1981.
124. D. S. Mileti and J. H. Sorensen, "Planning and Implementing Warning Systems," in *Mental Health Response to Mass Emergencies*, New York, NY, Bunner/Mazel, 1988, pp. 321–345.
125. G. Klein, Sources of Power: How People Make Decisions, Cambridge, MA: The MIT Press, 1999.
126. D. A. Gioia and P. P. Poole, "Scripts in Organizational Behavior," *The Academy of Management Review,* vol. 9, no. 3, pp. 449–459, July 1984.
127. P. Brennan, "Impact of Social Interaction to Time to Begin Evacuation in Office Buildings Fires: Implications for Modelling Behaviour," in *International Interflam Conference Seventh Proceedings*, London, England, 1996.
128. B. M. Levin, Ph.D., "Human Behavior in Fire: What We Know Now," Society of Fire Protection Engineers, Boston, MA, 1984.
129. J. D. Sime, "Affiliative Behaviour During Escape to Building Exits," *Journal of Environmental Psychology,* vol. 3, no. 1, pp. 21–41, March 1983.
130. D. Canter, J. Breaux and J. Sime, "Domestic, Multiple Occupancy, and Hospital Fires," *Fires and Human Behaviour,* pp. 117–136, 1980.
131. R. W. Bukowski, "Emergency Egress Strategies for Buildings," in *Interflam 2007, 11th Proceedings*, London, 2007.
132. E. D. Kuligowski and S. M. V. Gwynne, "The Need for Behavioral Theory in Evacuation Modeling," in *Proceedings of Pedestrian and Evacuation Dynamics*, Heidelberg, Germany, 2008.
133. E. D. Kuligowski, S. M. V. Gwynne, M. J. Kinsey and L. Hulse, "Guidance for the Model User on Representing Human Behavior in Egress Models," *Fire Technology,* pp. 1–24, 2016.
134. D. A. Purser and J. L. McAllister, "Assessment of Hazards to Occupants from Smoke, Toxic Gases, and Heat," in *The SFPE Handbook of Fire Protection Engineering*, 5 ed., vol. 3, New York, NY, Springer, 2016, pp. 2308–2428.
135. R. D. Stewart, "The Effects of Carbon Monoxide on Man," *Journal of Combustion Toxicology,* vol. 1, pp. 167–176, 1974.
136. R. R. Sayers and S. J. Davenport, "Review of Carbon Monoxide poisoning," United States Government Printing Office, Washington, D.C., 1930.
137. D. A. Purser and W. D. Woolley, "Biological Studies of Combustion Atmospheres," *Journal of Fire Sciences,* vol. 1, no. 2, pp. 118–144, March 1983.
138. D. A. Purser and K. R. Berrill, "Effects of Carbon Monoxide on Behaviour in Monkeys in Relation to Human Fire Hazard," *Archives of Environmental Health,* vol. 38, pp. 308–315, 1983.
139. D. A. Purser, "Chapter 4: Asphyxiant Components of Fire Effluents," *Fire Toxicity,* pp. 118–198, 2010.
140. G. Kimmerle, "Aspects and Methodology for the Evaluation of Toxicological Parameters during Fire Exposure," *Journal of Combustion Toxicology,* vol. 1, p. 4, 1974.
141. J. L. Bonsall, "Survival without Sequelae following Exposure to 500 mg/m3 of Hydrogen Cyanide," *Human & Experimental Toxicology,* vol. 3, no. 1, pp. 57–60, January 1984.
142. J. Barcroft, "The Toxicity Atmospheres Containing Hydrocyanic Acid Gas," *Journal of Hygiene,* vol. 31, no. 1, pp. 1–34, 1931.
143. B. P. McNamara, "Estimates of the Toxicity of Hydrocyanic Acid Vapors in Man," Edgewood Arsenal, Aberdeen Proving Ground, 1976.
144. D. A. Purser, "Chapter 3: Hazards from Smoke and Irritants," in *Fire Toxicity*, Cambridge, U.K., Woodhead Publishing, 2010, pp. 118–198.
145. D. A. Purser and P. Buckley, "Lung Irritation and Inflammation During and After Exposure to Thermal Decomposition Products from Polymeric Materials," *Medicine, Science and Law,* vol. 23, pp. 142–150, 1983.
146. D. A. Purser, "Application of Human Behaviour and Toxic Hazard Analysis to the Validation of CFD Modelling for the Mont Blanc Tunnel Fire Incident," in *Advanced Research Workshop: Fire Protection and Life Safety in Buildings and Transportation Systems*, University of Cantabria, Spain, 2009, pp. 30–31.
147. F. W. Beswick, P. Holland and K. H. Kemp, "Acute Effects of Exposure to Orthochlorobenzylidene Malononitrile (CS) and the Development of Tolerance," *British Journal of Industrial Medicine,* vol. 29, no. 3, pp. 298–306, July 1972.
148. D. L. Simms and P. L. Hinkley, "Fire Research Special Report No. 3: Protective Clothing Against Flames and Heat," Her Majesty's Stationary Office, 1960.
149. J. H. Veghte, "Fire Service Today," vol. 49, p. 16, 1982.

References

150. H. Elneil, "Man in Hot and Cold Environments," *A Companion to Medical Studies*, pp. 40.1–40.7, 1968.
151. C. S. Leithead and A. R. Lind, "Heat Stress and Heat Disorders," Cassel Publications, London, U.K., 1963.
152. K. Buettner, "Effects of Extreme Heat and Cold on Human Skin. II. Surface Temperature, Pain and Heat Conductivity in Experiments with Radiant Heat," *Journal of Applied Physiology*, vol. 3, no. 12, pp. 703–713, June 1951.
153. A. R. Moritz, F. C. Henriques and R. McLean, "The Effects of Inhaled Heat on the Air Passages and Lungs: An Experimental Investigation," *American Journal of Pathology*, pp. 311–331, 1945.
154. A. R. Moritz and F. C. Henriques, "Studies of Thermal Injury IV: An Exploration of Casualty-producing Attributes of Conflagrations. The Local and Systemic Effects of Generalized Cutaneous Exposure to Excessive Circumambient (air) and Circumambient Heat of Varying Duration and Intensity," *Archives of Pathology*, pp. 466–488, May 1947.
155. K. Kawagoe and H. Saito, "Measures to Deal with Smoke Problems Caused by Fire," *Journal of Japan Society for Safety Engineering*, vol. 6, no. 7, pp. 108–114, 1967.
156. D. J. Rasbash, "Smoke and Toxic Products Produced at Fires," *Plastics Institute Transaction and Journal*, pp. 55–61, January 1967.
157. F. E. Kingman, E. H. Coleman and D. J. Rasbash, "The Products of Combustion in Burning Buildings," *British Journal of Applied Chemistry*, vol. 3, pp. 463–468, 1953.
158. T. Jin, "Visibility through Fire Smoke - Part 5, Allowable Smoke Density for Escape from Fire," Fire Research Institute of Japan, 1976.
159. J. H. Shern, in *Sixth-Ninth Annual Meeting of the ASTM*, 1966.
160. J. A. Milke, D. E. Hugue, B. L. Hoskins and J. P. Carroll, "Tenability Analyses in Performance-Based Design," *Fire Protection Engineering Magazine*, 1 October 2005.
161. The Ministry of Business, "C/VM2 Verification Method: Framework for Fire Safety Design," Innovation and Employment (MBIE), Wellington, New Zealand, 2013.
162. National Fire Protection Association, NFPA 101[R] Life Safety Code[R], Quincy, MA, 2015.
163. International Standards Organization, "Fire Safety Engineering - Selection of Desgin Fire Scenarios and Design Fires," Geneva, Switzerland, 2015.
164. International Standards Organization, "Fire Safety Engineering - Selection of Design Occupant Behavioural Scenarios," Geneva, Switzerland, 2015.
165. Society of Fire Protection Engineers, "Engineering Guide: Fire Risk Assessment," Society of Fire Protection Engineers, Bethesda, MD, 2006.
166. S. M. V. Gwynne, E. D. Kuligowski and D. Nilsson, "Representing Evacuation Behaviour in Engineering Terms," *Journal of Fire Protection Engineering*, vol. 22, no. 2, pp. 133–150, 2012.
167. D. Purser, "Design Behavioral Scenarios for Escape Behavior Modeling in Tunnels and Underground Complexes," in *Advanced Research Workshop, Evacuation and Human Behavior in Emergency Situations*, University of Cantabria, Spain, 2011.
168. S. M. V. Gwynne, D. A. Purser, D. L. Boswell and A. Sekizawa, "Understanding and Representing Staff Pre-Warning Delay," *Journal of Fire Protection Engineering*, vol. 22, pp. 77–99, 2012.
169. International Standards Organization, "Life-threatening Components of Fire - Guidelines for the Estimation of Time to Compromised Tenability in Fires," Geneva, Switzerland, 2012.
170. D. A. Purser, "Effects of Pre-fire Age and Health Status on Vulnerability to Incapacitation and Death from Exposure to Carbon Monoxide and Smoke Irritants in Rosepark Fire Incident Victims," in *Sixth International Symposium Human Behaviour in Fire*, Cambridge, U.K., 2015.
171. G. E. Hartzell, H. W. Stacy, W. G. Switzer, D. N. Priest and S. C. Packham, "Modeling of Toxicological Effects of Fire Gases: IV. Intoxication of Rats by Carbon Monoxide in the Presence of a Toxicant," *Journal of Fire Science*, vol. 3, pp. 115–128, 1985.
172. U. C. Luft, "Aviation Physiology - The Effects of Altitude," in *Handbook of Physiology*, Washington, D.C., American Physiology Society, 1965, pp. 1099–1145.
173. International Standards Organization, "Fire-safety Engineering - Technical Information on Methods for Evaluating Behaviour and Movement of People," Geneva, Switzerland, 2009.
174. T. Jin, "Studies of Emotional Instability in Smoke from Fires," *Journal of Fire and Flammability*, vol. 12, pp. 130–142, April 1981.
175. H. Frantzich and D. Nilsson, "Evacuation Experiments in a Smoke Filled Tunnel," in *Third International Symposium on Human Behavior in Fire*, 2004.
176. V. Babrauskas, "Full Scale Burning Behavior of Upholstered Chairs," Technical Note 1103, National Bureau of Standards, Washington, DC, 1979.
177. K. Buettner, "Effects of Extreme Heat and Cold on Human Skin. II. Surface Temperature, Pain and Heat Conductivity in Experiments with Radiant Heat," *Journal of Applied Physiology*, vol. 3, pp. 703–713, 1951.
178. P. L. Simms and P. L. Hinkley, "Fire Research Special Report No. 3," Her Majesty's Stationary Office, London, 1963.
179. K. S. Mudan and P. A. Croce, "Thermal Radiation Model for LNG Trench Fires," in *ASME Winter Annual Meeting*, New Orleans, 1984.
180. T. Jin and T. Yamada, "Irritating Effects of Fire Smoke on Visibility," Fire Science & Technology, 1985.
181. T. Jin, "Visibility through Fire Smoke," Journal of Fire & Flammability, 1978, 1978.
182. G. Jensen, "Römming i røyk: Fullskala test av ledesystemer, personlig røykvern og atferd," IGP AS, Trondheim, 1993.
183. G. Jensen, "Evacuating in Smoke: Full Scale Tests on Emergency Egress Information Systems and Human Behaviour in Smoky Conditions," IGP AS, Trondheim, 1993.
184. T. Paulsen, "The Effect of Escape Route Information on Mobility and Way Finding Under Smoke Logged Conditions," in *Fire Safety Science*, Ottawa, 1994.
185. A. Tanaka, H. Imaizumi, S. Takahashi, T. Komai and T. Isei, "Evacuation from Underground Opening Space to Surface: Effect of Smoke," in *seventh International Fire Science and Engineering Conference*, Cambridge, 1996.
186. A. W. Heskestad and K. Schmidt Pedersen, "Escape Through Smoke: Assessment of Human Behaviour and Performance of Wayguidance Systems," in *The First International Symposium on Human Behaviour in Fire*, Ulster, 1998.
187. E. W. Janse, P. H. E. van de Leur and N. J. van Oerlo, "Evacuation from smoke filled corridors," in *the First International Symposium on Human Behaviour in Fire*, Ulster, 1998.
188. A. W. Heskestad, "Performance in Smoke of Wayguidance Systems," *Fire and Materials*, pp. 375–381, 1999.

189. H. Frantzich, "Utrymning av tunnelbanetåg: Experimentell utvärdering av möjligheten att utrymma i spårtunnel," Räddningsverket, Karlstad, 2000.
190. M. Wright, G. Cook and G. Webber, "The Effects of Smoke on People's Walking Speeds Using Overhead Lighting and Wayduidance Provision," in *The second International Symposium on Human Behaviour in Fire*, Ulster, 2001.
191. H. Frantzich and D. Nilsson, Utrymning genom tät rök: beteende och förflyttning, Lund: Lunds universitet, 2003.
192. H. Frantzich and D. Nilsson, "Evacuation Experiments in a Smoke Filled Tunnel," in *Third International Symposium on Human Behaviour in Fire*, London, 2004.
193. Y. Akizuki, K. Yamao and T. Tanaka, "Experimental Study On Walking Speed In Escape Route Considering Luminous Condition, Smoke Density And Evacuee's Visual Acuity," in *The 7th Asia-Oceania Symposium on Fire Science and Technology*, 2007.
194. H. Xie, L. Filippidis, E. Galea, D. Blackshields and P. Lawrence, "Experimental study of the effectiveness of emergency signage," in *fourth International Symposium on Human Behaviour in Fire*, Cambridge, 2009.
195. K. Fridolf, E. Ronchi, D. Nilsson and H. Frantzich, "Movement speed and exit choice in smoke-filled rail tunnels," *Fire Safety Journal,* vol. 59, pp. 8–21, 2013.
196. K. Fridolf and H. Frantzich, "Test av vägledande system i en tunnel," Lunds universitet, Lund, 2015.
197. M. Seike, N. Kawabata and M. Hasegawa, "Experiments of evacuation speed in smoke-filled tunnel," *Tunnelling and Underground Space Technology,* vol. 53, pp. 61–67, 2016.
198. K. Fridolf, D. Nilsson, H. Frantzich, E. Ronchi and S. Arias, "Människors gånghastighet i rök: Förslag till representation vid brandteknisk projektering," SP Sveriges Tekniska Forskningsinstitut, Borås, 2016.
199. K. Fridolf, D. Nilsson, H. Frantzich, E. Ronchi and S. Arias, "Walking Speed in Smoke: Representation in Life Safety Verifications," in *the 12th International Performance-Based Codes and Fire Safety Design Methods Conference*, Oahu, Hawaii, 2018.
200. J. Pauls, "Building Evacuation: Research Findings and Recommendations," in *Fires and Human Behaviour*, D. Canter, Ed., New York, NY, John Wiley, 1980.
201. J. J. Fruin, Pedestrian Planning and Design, revised edition, Mobile, Alabama : Elevator World Educational Services Division, 1987.
202. A. Habicht and J. Braaksma, "Effective Width of Pedestrian Corridors," *Journal of Transportation Engineering,* vol. 110, no. 1, pp. 80–93, January 1984.
203. S. M. V. Gwynne and E. R. Rosenbaum, "Employing the Hydraulic Model in Assessing Emergency Movement," in *SFPE Handbook of Fire Protection Engineering, 5th edition*, Chapter 59, Gaithersburg, Maryland: Society of Fire Protection Engineers, 2016, pp. 2115–2151.
204. National Fire Protection Association, NFPA 101® Life Safety Code®, Quincy, MA, 2000.
205. B. L. Hoskins and J. A. Milke, "Differences in Measurement Method for Travel Distance and Area for Estimates of Occupant Speed on Stairs," *Fire Safety Journal,* vol. 48, pp. 49–57, 2012.
206. G. Proulx, "Evacuation Time and Movement in Apartment Buildings," *Fire Safety Journal,* vol. 24, no. 3, pp. 229–246, December 1994.
207. A. P. Adams and E. R. Galea, "An Experimental Evaluation of Movement Devices Used to Assist People with Reduced Mobility in High-Rise Building Evacuations," in *Pedestrian and Evacuation Dynamics*, R. D. Peacock, E. D. Kuligowski and J. D. Averill, Eds., New York, Springer, 2011, pp. 129–138.
208. E. D. Kuligowski, R. Peacock, E. Wiess and B. L. Hoskins, "Stiar Evacuation of Older Adults and People with Mobility Impairments," *Fire Safety Journal,* no. 62, pp. 230–237, 2013.
209. E. D. Kuligowski, R. Peacock, E. Wiess and B. L. Hoskins, "Stair Evacuation of People with Mobility Impairments," *Fire and Materials,* vol. 39, no. 4, pp. 371–384, June 2015.
210. V. M. Predtechenskiĭ and A. I. Milinskiĭ, Planning for Foot Traffic Flow in Buildings, New Delhi: Amerind Publishing Co., 1978.
211. H. Frantzich, "Study of Movement on Stairs During Evacuation using Video Analysing Techniques," Lund Institute of Technology, Lund, Sweden, 1996.
212. B. L. Hoskins, "Adjusted Density Measurement Methods on Stairs," *Fire and Materials,* vol. 39, no. 4, pp. 323–334, June 2015.
213. T. J. Shields, K. Dunlop and G. Silcock, "Escape of Disabled People from Fire - A Measurement and Classification of Capability for Assessing Escape Risk," British Research Establishment, Borehamwood, U.K., 1996.
214. A. R. Larusdottir, A. Dederichs and D. Nilsson, "Evacuation of Children: Focusing on Daycare Centers and Elementary Schools," Technical University of Denmark, Department of Civil Engineering, 2013.
215. J. A. Templer, "Stair Shape and Human Movement," Columbia University, 1975.
216. Y. Murosaki, H. Hayashi and T. Nishigaki, "Effects of Passage Width on Choice of Egress Route at a T-Junction in a Building," in *First International Symposium on Fire Safety Science*, Hemisphere, NY, 1986.
217. E. D. Kuligowski and S. M. V. Gwynne, "What a User Should Know When Selecting an Evacuation Model," *Fire Protection Engineering,* no. 28, pp. 30–35, 2005.
218. S. M. Strege and M. J. Ferreira, "Modeling Occupant Ingress at a Secure Building and Using Field Measurements for Calibration," *Transportation Research Procedia,* pp. 807–812, 2 October 2014.
219. S. M. Strege and S. Goodhead, "People Movement Study of Large Airport -- Data Generation, Flow Dynamics and Coupled Analysis," in *6th International Symposium on Human Behavior in Fire*, Cambridge, UK, 2015.
220. E. R. Galea, G. Sharp and P. J. Lawrence, "Investigating the Representation of Merging Behavior at the Floor-Stair Interface in Computer Simulations of Multi-Floor Building Evacuations," *Journal of Fire Protection Engineering,* vol. 04, March 2008.
221. K. E. Boyce, D. A. Purser and J. T. Shields, "Experimental Studies to Investigate Merging Behaviour in a Staircase," *Fire and Materials,* vol. 36, no. 5–6, pp. 383–398, March 2011.
222. E. D. Kuligowski and R. D. Peacock, "A Review of Building Evacuation Models," National Institute of Standards and Technology, Gaithersburg, MD, 2005.
223. E. D. Kuligowski, R. D. Peacock and B. L. Hoskins, "A Review of Building Evacuation Models, 2nd Edition (Rep. No. NIST TN 1680)," NIST, Gaithersburg, MD, 2010.
224. The American Institute of Aeronautics and Astronautics, "AIAA Guide for the Verification and Validation of Computational Fluid Dynamics Simulations," Reston, VA, 1998.

References

225. J. Lord, B. J. Meacham and A. Moore, "Guide for Evaluating the Predictive Capabilities of Computer Egress Models," National Institute of Standards and Technology, Gaithersburg, MD, 2005.
226. M. Richiardi, R. Leombruni, N. Saam and M. Sonnessa, "A Common Protocol for Agent-Based Social Simulation," *Journal of Artificial Societies and Social Simulation,* vol. 9, no. 1, January 2006.
227. R. J. Malak and C. J. Paredis, "Foundations of Validating Reusable Behavioral Models in Engineering Design Problems," in *Proceedings of the 2004 Winter Simulation Conference*, 2004.
228. G. K. Bharathy and B. G. Silverman, "Validating Agent Based Social Systems Models," in *2010 Winter Simulation Conference*, Baltimore, MD, 2010.
229. R. G. Sargent, "Verification and Validation of Simulation Models," in *Proceedings of the 2010 Winter Simulation Conference*, Baltimore, MD, 2010.
230. O. Balci, R. E. Nance, J. D. Arthur and W. F. Ormsby, "Expanding our Horizons in Verification, Validation and Accreditation Research and Practice," in *Winter Simulation Conference*, Blacksburg, VA, 2002.
231. R. G. Sargent, "A New Statistical Procedure for Validation of Simulation and Stochastic Models," Syracuse University, Syracuse, NY, 2010.
232. R. G. Sargent, "Some Subjective Validation Methods using Graphical Displays of Data," in *Proceedings of the 1996 Winter Simulation Conference*, 1996.
233. R. G. Sargent, "A Tutorial on Verification and Validation of Simulation Models," in *Proceedings of the 1984 Winter Simulation Conference*, 1984.
234. R. G. Sargent, "Some Approaches and Paradigms for Verifying and Validating Simulation Models," in *Proceedings of the 2001 Winter Simulation Conference*, 2001.
235. J. P. Kleijnen, "Theory and Methodology: Verification and Validation of Simulation Models," *European Journal of Operation Research,* vol. 82, pp. 145–162, 1995.
236. M. D. Petty, "Calculating and Using Confidence Intervals for Model Validation," in *Proceedings of the Fall 2012 Simulation Interoperability Workshop*, Orlando, FL, 2012.
237. W. J. White, A. Rassweiler, J. F. Samhouri, A. C. Stier and C. White, "Ecologists Should Not Use Statistical Significance Tests to Interpret Simulation Model Results," Oikos, 2013.
238. M. D. Petty, "Advanced Topics in Calculating and Using Confidence Intervals for Model Validation," in *Proceedings of the Spring 2013 Simulation Interoperability Workshop*, San Diego, CA, 2013.
239. M. D. Petty, "Modeling and Validation Challenges for Complex Systems," in *The Proceedings of the Spring 2012 Simulation Interoperability Workshop*, Orlando, FL, 2012.
240. M. D. Petty, "Verification, Validation, and Accreditation," in *Modeling and Simulation Fundamentals: Theoretical Underpinnings and Practical Domains*, J. A. Sokolowski and C. M. Banks, Eds., Hoboken, NJ: John Wiley & Sons, Inc., 2010, pp. 325–372.
241. W. E. Baker and M. S. Taylor, "A Nonparametric Statistical Approach to the Validation of Computer Simulation Models," US Army Ballistic Research Laboratory, Aberdeen Proving Ground, MD, 1985.
242. M. K. Dey, "Lessons Learned in the ICFMP Project for Verification and Validation of Computer Models for Nuclear Plant Fire Safety Analysis," in *Ninth International Conference on Performance-Based Codes and Fire Safety Design Methods*, Hong Kong, 2012.
243. W. K. Mok and W. K. Chow, ""Verification and Validation" in Modeling Fire by Computational Fluid Dynamics," *International Journal on Architectural Science,* vol. 5, no. 3, pp. 58–67, 2004.
244. G. Rein, E. Abecassis and R. Carvel, Eds., The Dalmarnock Fire Tests: Experiments and Modelling, University of Edinburgh: School of Engineering and Electronics, 2007.
245. C. Abecassis-Empis, P. Reszka, A. Cowlard, H. Biteau, S. Welch, G. Rein and J. L. Torero, "Characterization of Dalmarnock Fire Test One," *Experimental Thermal and Fluid Science,* vol. 32, no. 7, pp. 1334–1343, 2008.
246. W. Jahn, G. Rein and J. L. Torero, "The Effect of Model Parameters on the Simulation of Fire Dynamics," *Fire Safety Science,* vol. 9, pp. 1341–1352, 2008.
247. G. Rein, J. L. Torero, W. Jajn, J. Stern-Gottfried, N. L. Ryder, S. Desanghere, M. Làzaro, F. Mowrer, A. Coles, D. Joyeux, D. Alvear, J. A. Capote, A. Jowsey, C. Abecassis-Empis and P. Reszka, "Round-Robin Study of a Prior Modelling Predictions of The Dalmarnock Fire Test One," *Fire Safety Journal,* vol. 44, no. 4, pp. 590–602, 2009.
248. E. Ronchi, E. D. Kuligowski, P. A. Reneke, R. D. Peacock and D. Nilsson, "The Process of Verification and Validation of Building Fire Evacuation Models," National Institute of Standards and Technology, Gaithersburg, MD, 2013.
249. E. R. Galea, The Validation of Evacuation Models, CMS Press, 1997.
250. M. Kinsey, G. Walker, N. Swailes and N. Butterworth, "MassMotion: The Verification and Validation of MassMotion for Evacuation Modelling," Arup, 2015.
251. International Maritime Organization, "Guidelines for Evacuation Analysis for New and Existing Passenger Ships," London, U.K., 2007.
252. W. Radax and B. Rengs, "Prospects and Pitfalls of Statistical Testing: Insights from Replicating the Demographic Prisoner's Dilemma," *Journal of Artificial Societies and Social Simulation,* vol. 13, no. 4, p. 1, 2010.
253. International Standards Organization, "Fire Safety Engineering - Assessment, Verification and Validation of Calculation Methods," Geneva, Switzerland, 2008.
254. R. L. Best, "Tragedy in Kentucky," *Fire Journal,* vol. 72, no. 1, pp. 41–44, 1978.
255. G. Proulx, "Occupant Behavior and Evacuation during the Chicago Cook County Administration Building Fire," *Journal of Fire Protection Engineering,* vol. 16, no. 4, pp. 283–309, 2006.
256. H. Jiang, S. M. V. Gwynne, E. Galea, P. Lawrence, F. Jia and I. H, "The Use of Evacuation Simulation, Fire Simulation and Experimental Fire Data in Forensic Fire Analysis," in *Second Conference on Pedestrian and Evacuation Dynamics*, University of Greenwich, U.K., 2003.
257. D. A. Purser, "Fire Safety and Evacuation Implications from Behaviours and Hazard Development in Two Fatal Care Home Incidents," *Fire and Materials,* vol. 39, no. 4, pp. 430–452, June 2015.
258. J. L. Bryan, "Human Behavior in the Westchase Hilton Hotel Fire," *Fire Journal,* vol. 77, pp. 78–85, 1983.

259. J. Parke, S. M. V. Gwynne, E. R. Galea and P. Lawrence, "Validating the Building EXODUS Evacuation Model using Data from a Unannounced Trial Evacuation," in *Proceeding of the Second International Conference on Pedestrian and Evacuation Dynamics*, University of Greenwich, U.K., 2003.

260. S. M. V. Gwynne and D. L. Boswell, "Pre-Evacuation Data Collected from a Mid-Rise Evacuation Exercise," *Journal of Fire Protection Engineering,* vol. 19, no. 1, pp. 5–29, 2009.

261. G. Proulx, "Lessons Learned On Occupants' Movement Times and Behaviour During Evacuation Drills," in *Proceedings from the Seventh International Interflam Conference*, London, U.K., 1996.

262. S. Schmidt and E. Galea, Behaviour - Security - Culture Human Behaviour in Emergencies and Disasters: A Cross-Cultural Investigation, Pabst Science Publishers, 2013, p. 256.

263. J. P. A. Loannidis, "Why Most Published Research Findings are False," *PLoS Medicine,* vol. 2, no. 8, pp. 696–701, August 2005.

264. J. A. Kopec, P. Finès, D. G. Manuel, D. L. Buckeridge, W. M. Flanagan, J. Oderkirk, M. Abrahamowicz, S. Harper, B. Sharif, A. Okhmatovskaia, E. C. Sayre, M. M. Rahman and M. C. Wolfson, "Validation of Population-Based Disease Simulation Models: A Review of Concepts and Methods," *BMC Public Health,* 2010.

265. A. H. Feinstein and H. M. Cannon, "Fidelity, Verifiability, and Validity of Simulation: Constructs for Evaluation," *Developments in Business Simulation and Experiential Learning,* vol. 28, pp. 57–67, 2001.

266. E. McCrum-Gardner, "Which is the Correct Statistical Test to Use?," *British Journal of Oral and Maxillofacial Surgery,* vol. 46, no. 1, pp. 38–41, 2008.

267. O. Balci and R. G. Sargent, "A Methodology for Cost-risk Analysis in the Statistical Validation of Simulation Models," *Communications of the ACM,* vol. 24, no. 4, pp. 190–197, 1 April 1981.

268. I. I. Sandra, "Model's Validation for Complex Real-Time Systems," Department of Computer Science and Engineering, Västerås, Sweden, 2004.

269. E. R. Galea, L. Filippidis, S. Deere, R. Brown, I. Nicholls, Y. Hifi and N. Besnard, "The SAFEGUARD Validation Data-Set and Recommendations to IMO to Update MSC/Circ. 1238," in *Safeguard Passenger Evacuation Seminar*, London, UK, 2012.

270. R. D. Peacock, P. A. Reneke, W. D. Davis and W. W. Jones, "Quantifying Fire Model Evaluation using Functional Analysis," *Fire Safety Journal,* vol. 33, pp. 167–184, 1999.

271. E. R. Galea, S. Deere, R. Brown and L. Filippidis, "An Experimental Validation of an Evacuation Model using Data Sets Generated from Two Large Passenger Ships," in *Pedestrian and Evacuation Dynamics 2012*, U. Weidmann, U. Kirsch and M. Schreckenberg, Eds., Springer International Publishing, 2013, pp. 109–123.

272. A. M. Law, "How to Build Valid and Credible Simulation Models," in *Proceedings of the 2009 Winder Simulation Conference*, 2009.

273. D. A. Cook and J. A. Skinner, "How to Perform Credible Verification, Validation, and Accreditation for Modeling and Simulation," *The Journal of Defense Software Engineering,* pp. 20–24, 2005.

274. Society of Fire Protection Engineers, *Guidelines for Substantiating a Fire Model for a Given Application,* Bethesda, Maryland, 2010.

275. A. Hamins and K. B. McGrattan, "Verification & Validation of Selected Fire Models for Nuclear Power Plant Applications," U.S. Nuclear Regulatory Commission, Rockville, MD, 2007.

276. J. D. Averill and W. Song, "Accounting for Emergency Response in Building Evacuations: Modeling Differential Egress Capacity Solutions," National Institute of Standards and Technology, 2007.

277. Standard Norge, "prINSTA TS 950 Fire Safety Engineering — Verification of fire safety design in buildings," 2013.

278. E. D. Kuligowski, S. M. V. Gwynne, K. M. Butler, B. L. Hoskins and C. R. Sandler, "Developing Emergency Communication Strategies for Buildings," National Institute of Standards and Technology, Gaithersburg, MD, 2012.

279. E. D. Kuligowski and H. Omori, "General Guidance on Emergency Communication Strategies for Buildings, Second Edition," National Institute of Standards and Technology, Gaithersburg, MD, 2014.

280. U.S. Department of Justice, ADA Standards for Accessible Design, Washington, D.C.: U.S. Department of Justice, 2010.

281. L. Campbell, C. Carney and B. H. Kantowitz, "Human Factors Design Guidelines for Advanced Traveler Information Systems (ATIS)and Commercial Vehicle Operations (CVO)," U.S. Department of Transportation Federal Highway Administration, Washington, DC, Report #FHWA–RD–98–057.

282. National Fire Protection Association, National Fire Alarm Code and Signaling Handbook, Quincy, MA: National Fire Protection Association, 2016.

283. B. S. E. Norm, *BS EN 60849,* London: British Standard European Norm, 1998.

284. N. Groner, "A Simple Decision Model for Managing the Movement of Building Occupants during Fire Emergencies," in *Proceedings from the Fifth International Symposium on Human Behaviour in Fire*, London, U.K., 2015.

285. N. Groner, "A Decision Model for Recommending which Building Occupants should Move Where During Fire Emergencies," *Fire Safety Journal,* vol. 80, pp. 20–29, February 2016.

286. J. M. Chertkoff and R. H. Kushigian, Don't Panic: The Psychology of Emergency Egress and Ingress, Westport, CT: Praeger, 1999.

287. J. Burtles, Emergency Evacuation Planning: From Chaos to Life-Saving Solutions, Brookfield, CT: Rothstein, 2013.

288. UK.gov, "Fire safety advice documents," 30 May 2013. [Online]. Available: https://www.gov.uk/workplace-fire-safety-your-responsibilities/fire-safety-advice-documents. [Accessed 15 July 2013].

289. J. S. Tubbs and B. J. Meacham, Egress Design Solutions: A Guide to Evacuation and Crowd Management Planning, Hoboken, NJ: John Wiley & Sons, 2007.

290. R. W. Bukowski and J. S. Tubbs, "Chapter 56, Egress Concepts and Design. SFPE Handbook of Fire Protection Engineering, 5th edition," SFPE, Gaithersburg, MD, 2016.

291. G. Proulx, "High-rise Evacuation: A Questionable Concept," in *Proceedings of the Second International Symposium on Human Behaviour in Fire*, London, U.K., 2001.

292. J. Burtles, Emergency Evacuation Planning, Brookfield, CT: Rothstein, 2013.

293. NFPA, "Guide to the Fire Safety Concepts Tree, NFPA 550," NFPA, Quincy, MA, 2017.

294. N. E. Groner, "On Not Putting the Cart before the Horse: Design Enables the Prediction of Decisions about Movement in Buildings," 2005, pp. 96–98.

References 159

295. R. F. Duval, "Fire Investigations: Nursing Home; Harford, CT; February 26, 2003," National Fire Protection Association, Quincy, MA, 2005.
296. Society of Fire Protection Engineers, Engineering Guide: Fire Safety for Very Tall Buildings, Bethesda, MD: Society of Fire Protection Engineers, International Code Council, 2013.
297. D. O'Conner, K. Clawson and E. Cui, "Consideration and Challenges for Refuge Areas in Tall Buildings," in *CTBUH 9th International Congress*, Shanghia, 2012.
298. J. J. O'Donoghue, "Evolution of an Elevator Task Group Project," *Fire Engineering,* vol. 163, no. 3, March 2010.
299. R. D. Peacock, "Summary of NIST/GSA Cooperative Research on the Use of Elevators During Fire Emergencies," National Institute of Standards and Technology, Gaithersburg, MD, 2009.
300. G. E. Hedman, "Status report on the development of the RESNA performance standard for emergency stair travel devices," in *Human Behavior in Fire - 5th International Symposium*, Cambridge, UK, 2012.
301. RESNA, "RESNA-ED-1: American National Standard for Evacuation Devices - Vol. 1," Rehabilitation Engineering and Assistive Technology Society of North America, Arlington, VA, 2013.
302. A. B. Fraser, "Emergency Evacuation Planning Guide for People with Disabilities," National Fire Protection Association, Quincy, MA, 2007.
303. National Organization on Disability, "Functional Needs of People with Disabilities: A Guide for Emergency Managers, Planners and Responders," 2009. [Online]. Available: http://nod.org/disability_resources/emergency_preparedness_for_persons_with_disabilities/. [Accessed 14 January 2014].
304. J. R. Hall, Jr, "U.S. Experience with Sprinklers," National Fire Protection Association, Quincy, MA, 2013.
305. N. Groner, "Can the cognitive engineering approach prevent "normal accidents"? How design might improve societal resiliency to critical incidents," *Journal of Critical Incident Analysis,* vol. 1, no. 2, pp. 96–104, 2011.
306. C. Perrow, Normal Accidents: Living with High Risk Technologies, Princeton, NJ: Princeton University Press, 1999.
307. M. R. Endsley, B. Bolte and D. G. Jones, Designing for Situation Awareness: An Approach to User-Centered Design, Bocva Raton, FL: CRC Press, 2003.
308. N. E. Groner, "Situation Awareness Requirements Analysis For Emergency Management Planning (Working Paper 90–02)," Christian Regenhard Center for Emergency Response Studies, New York, 2009.
309. L. G. Perry, Are your Tenants Safe? BOMA's Guide to Security and Emergency Planning, Washington, D.C.: BOMA International, 2000.
310. Council on Tall Buildings in Urban Habitat, Building Safety Assessment Guidebook, Chicago, IL: Council on Tall Buildings in Urban Habitat, 2002.
311. L. J. Shoen, Emergency Preparedness Guidebook: The Property Professional's Resource for Developing Emergency Plan for Natural and Human-Based Threats, Washington, DC: BOMA International, 2012.
312. J.R. Hall, Jr. Directions and Strategies for Research on Human Behavior and Fire, Are We Prepared to Support Decision-Making on the Major Themes?, (2004), The Third International Symposium on Human Behavior in fire, London.
313. T. Jin, Visibility and Human Behavior in Smoke. SFPE Handbook of Fire Protection Engineering, 3rd edition. National Fire Protection Association. Quincy, MA 2002.
314. K. A. Notarianni and G. W. Parry, "Uncertainty," in SFPE Handbook of Fire Protection Engineering, 5th edition, Chapter 76, Gaithersburg, Maryland: Society of Fire Protection Engineers, 2016.

Printed in the United States
By Bookmasters